WHALE FISHERY OF NEW ENGLAND

AN ACCOUNT,
WITH ILLUSTRATIONS AND SOME INTERESTING AND AMUSING ANECDOTES, OF THE RISE AND FALL OF AN INDUSTRY WHICH HAS MADE NEW ENGLAND FAMOUS THROUGHOUT THE WORLD

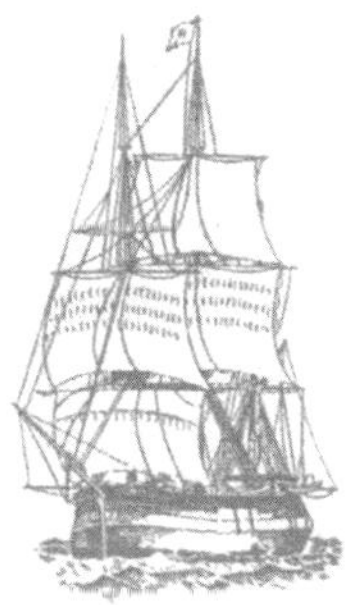

PRINTED FOR THE
STATE STREET TRUST COMPANY
BOSTON, MASS.

Marula

An imprint of Arcadia Publishing

Whale Fishery of New England
An Account, with Illustrations and Some Interesting and Amusing Anecdotes, of the Rise and Fall of an Industry Which Has Made New England Famous throughout the World

State Street Trust Company

A Marula Historical Reprint
Originally published in 1915

Published by Arcadia Publishing
Charleston SC

ISBN 978-0-7385-9492-7

Printed in the United States of America

For all general information contact Arcadia Publishing at:
Telephone 843-853-2070
Fax 843-853-0044
E-mail sales@arcadiapublishing.com
For customer service and orders:
Toll-Free 1-888-313-2665

Visit us on the Internet at www.arcadiapublishing.com

The vignette on the title-page is reproduced from a print of the ship "Maria" of New Bedford, which in 1853 was the oldest whaleship owned in the United States. Her registry was dated 1782. She was built in Pembroke, now called Hanson, for a privateer during the Revolutionary War, and was bought in the year 1783 by William Rotch of Nantucket, afterwards of New Bedford. At one time she was owned by Samuel Rodman, and also by the Russells. In construction she was the typical whaleship of her time. It is said that she earned for her owners $250,000 and made twenty-five voyages, bringing back a full cargo each time. The tailpiece is from a very old print which represents whaling in the seventeenth century.

Compiled, arranged and printed
under the direction of the
Walton Advertising and Printing Company
Boston, Mass.

FOREWORD

The people of New England have long been interested in all matters pertaining to the sea, and members of many of her best-known families have commanded its merchant ships and whalers.

The State Street Trust Company has always endeavored to encourage an interest in historical matters, and it is hoped that this pamphlet, the ninth of the series, which deals with one of our earliest industries, will be interesting to the Company's depositors and also to the general public. It is sent to you with the compliments of the Company, which for over twenty years has tried to serve the interests of its depositors.

For valuable assistance in the preparation of this pamphlet the Trust Company desires to acknowledge its indebtedness to Dr. Benjamin Sharp and Sidney Chase, residents of Nantucket (the latter being a descendant of the Starbucks, Coffins and Husseys), to Z. W. Pease, Frank Wood and George H. Tripp, all of New Bedford (Mr. Tripp being the librarian of the Free Public Library), Llewellyn Howland, Frederick P. Fish, Charles H. Taylor, Jr., Roy C. Andrews and Madison Grant of the American Museum of Natural History, New York, D. A. deMenocal, J. E. Lodge and Kojiro Tornita of the Museum of Fine Arts, Boston, and George F. Lord, secretary of the Boston Stock Exchange. Assistance has also been rendered by the officers of the Trust Company.

The following books have been used as references and contain valuable information and many interesting anecdotes:—

"The Story of New England Whalers," by John R. Spears.
"History of the American Whale Fishery," by Alexander Starbuck.
"A History of the American Whale Fishery," by Walter S. Tower.
"Moby Dick, or the White Whale," by Herman Melville.
"Whaling Ventures and Adventures," by George H. Tripp.
"Whaling and Fishing," by Charles Nordhoff.
"Miriam Coffin," by Col. Joseph C. Hart.
"The Gam," by Capt. Charles Henry Robbins.
"Eighteen Months on a Greenland Whaler," by Joseph P. Faulkner.
"Arctic Whaleman and Whaling," by Rev. Lewis Holmes.
"Cruise of the Cachalot," by Frank T. Bullen.
"History of Nantucket," by Edward K. Godfrey.
"History of Nantucket," by Obed Macy.
"History of Nantucket," by Douglas-Lithgow.
"The Glacier's Gift" (Nantucket), by Eva C. G. Folger.
"History of New Bedford," by Daniel Ricketson.
"The Perils and Romance of Whaling," by G. Kobbé.
"The Whale and its Captors," by Rev. Henry T. Cheever.
"Incidents of a Whaling Voyage," by Olmstead.
"Nimrod of the Sea," by Captain Davis.
"Hunting the Biggest of all Big Game," by Roy C. Andrews.
"Four Years Aboard a Whaleship," by William B. Whitecar, Jr.
"Etchings of a Whaling Cruise," by J. Ross Browne.
"Bark Kathleen, sunk by a Whale," by Capt. T. H. Jenkins.
"Peter the Whaler," by William H. G. Kingston.
"The Fisheries and Fishery Industries of the United States,"
by George Brown Goode, prepared for the United States Tenth Census.

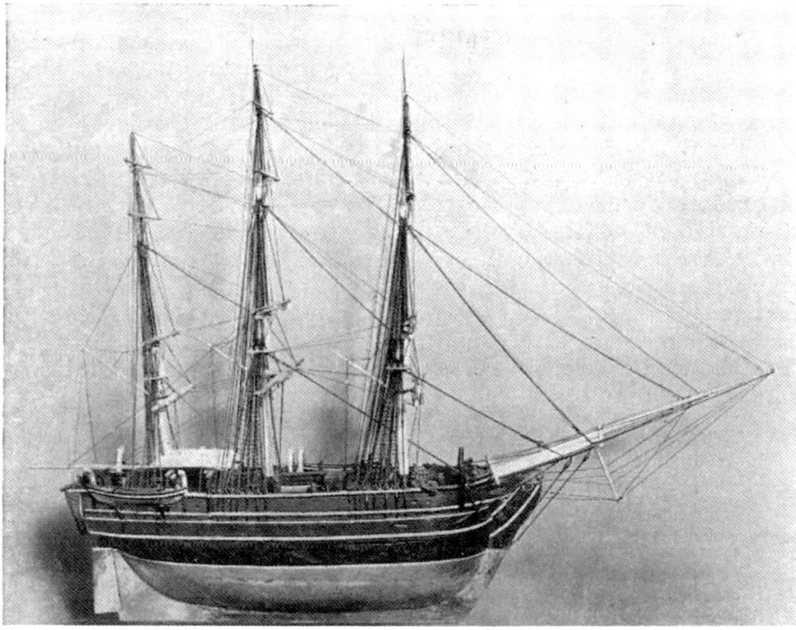

Model of the whaleship "Henry," made at sea in 1847. This model stands in the main banking rooms of the Company, and may be seen by visitors.

CONTENTS

The illustrations used in this brochure are from rare prints in the possession of the Dartmouth Historical Society and the Free Public Library of New Bedford, H. S. Hutchinson & Co., Charles H. Taylor, Jr., Roy C. Andrews of the American Museum of Natural History of New York, Doubleday, Page & Co., and others.

"Neither the perseverance of Holland, nor the activity of France, nor the dexterous and firm sagacity of the English enterprise, ever carried this most perilous mode of hardy industry to the extent to which it has been pushed by this recent People; a People who are still, as it were, but in the gristle, and not yet hardened into the bone, of manhood."—*From a speech by Edmund Burke before Parliament in 1775.*

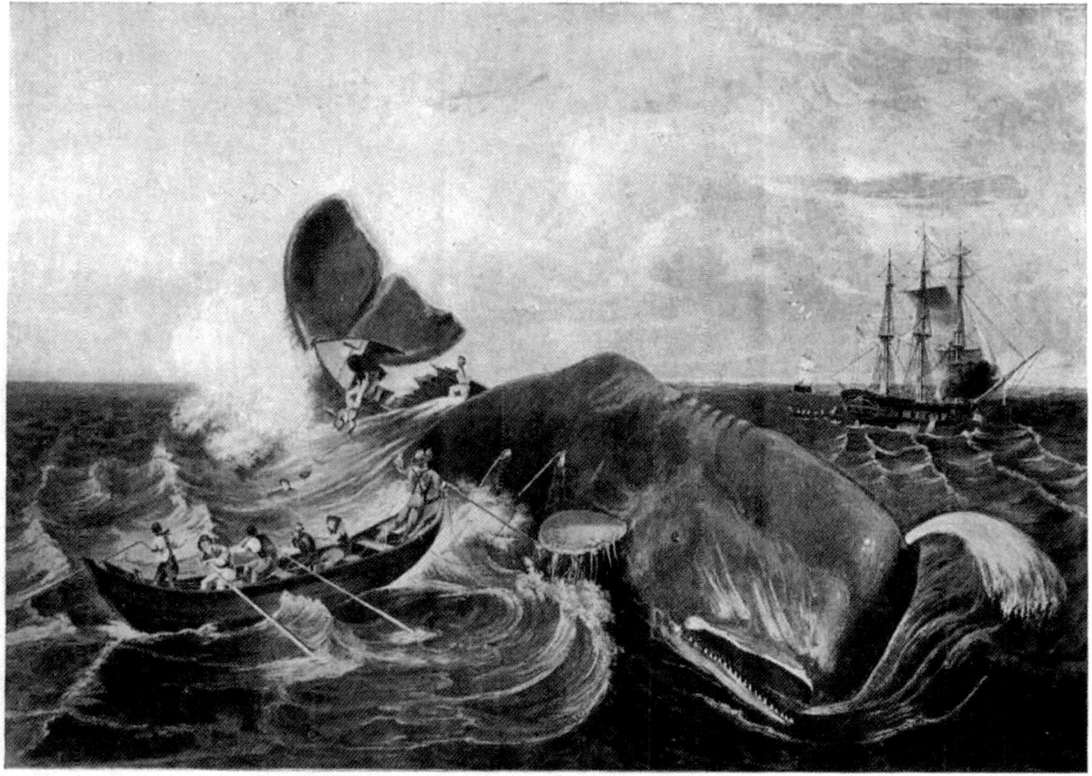

Capturing a huge sperm whale. (From a very rare print.)

THE WHALE

"Oh, the rare old Whale, 'mid storm and gale,
In his ocean home will be
A giant in might where might is right,
And King of the boundless sea."

From "Moby Dick."

NO ANIMAL in prehistoric or historic times has ever exceeded the whale, in either size or strength, which explains perhaps its survival from ancient times. Few people have any idea of the relative size of the whale compared with other animals. A large specimen weighs about ninety tons, or thirty times as much as an elephant, which beside a whale appears about as large as a dog compared to an elephant. It is equivalent in bulk to one hundred oxen, and outweighs a village of one thousand people. If cut into steaks and eaten, as in Japan, it would supply a meal to an army of one hundred and twenty thousand men.

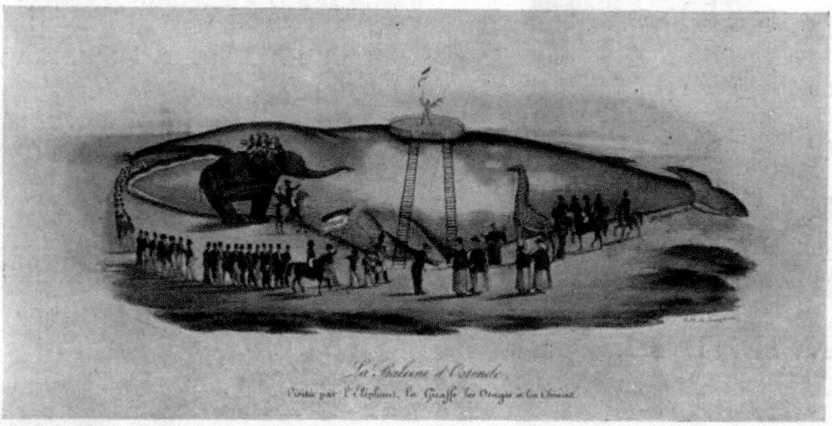

A French lithograph showing the comparative sizes of a whale, an elephant a horse, and a giraffe.

Whales have often exceeded one hundred feet in length, and George Brown Goode, in his report on the United States Fisheries, mentions a finback having been killed that was one hundred and twenty feet long. A whale's head is sometimes thirty-five feet in circumference, weighs thirty tons, and has jaws twenty feet long, which open thirty feet wide to a mouth that is as large as a room twenty feet long, fifteen feet high, nine feet wide at the bottom, and two feet wide at the top. A score of Jonahs standing upright would not have been unduly crowded in such a chamber.

The heart of a whale is the size of a hogshead. The main blood artery is a foot in diameter, and ten to fifteen gallons of blood pour out at every pulsation. The tongue of a right whale is equal in weight to ten oxen, while the eye of all whales is hardly as large as a cow's, and is placed so far back that it has in direction but a limited range of vision. The ear is so small that it is difficult to insert a knitting needle, and the brain is only about ten inches square. The head, or "case," contains about five hundred barrels, of ten gallons each, of the richest kind of oil, called spermaceti.

One of these giants, when first struck by a harpoon, can go as fast as a steam yacht, twenty or twenty-five miles an hour, but it soon slows down to its usual speed of about twelve miles, developing about one hundred and forty-five horse-power.

Mr. Roy C. Andrews, of the American Museum of Natural History, New York, was on a whaler ninety feet long, which struck a finback whale, and he says that for seven hours the whale towed the vessel, with engines going at full speed astern, almost as though it had been a rowboat.

The whale's young are about twelve feet long at birth, and can swim as soon as they are born. So faithfully does the cow whale watch over her offspring when they are together that she will rarely move when attacked for fear of leaving the young whale unprotected, or of hurting it if she thrashes round to escape capture. It is believed that whales sometimes live to attain the age of eight hundred years. They sleep at the bottom of the ocean, which fact shows that they do not inhale air when asleep, like the warm-blooded animals, and to help them in breathing below the surface they have a large reservoir of blood to assist circulation. This spot is known to whalemen as the "life" of the whale. When "sounding" to a great depth it is estimated that the whale bears on its back the weight of twenty battleships. The strength and power of a whale are described as almost unbelievable.

ANCIENT HISTORY OF WHALING

Every one knows the story of Jonah; how he was thrown overboard to appease the gods, and how a "big fish" swallowed him and carried him ashore. It will always be a mooted question whether or not the big fish was a whale. If it were a whale, it is doubtful whether Jonah got any further than its mouth, on account of the smallness of a whale's throat. It may be well to explain that a whale does not belong to the fish family, but is a mammal, and therefore, perhaps, this great fish mentioned wasn't a whale.

This "fishing on a gigantic scale," as it has been often termed, is of very ancient origin and dates back to 890 A.D., when a Norwegian, called Octhere, skirted the coast of Norway for whales.

The Biscayans, who in the twelfth, thirteenth, fourteenth, and fifteenth centuries became famous on account of their whale fishery, were the first people to prosecute this industry as a regular commercial pursuit. In this connection the French are also mentioned about

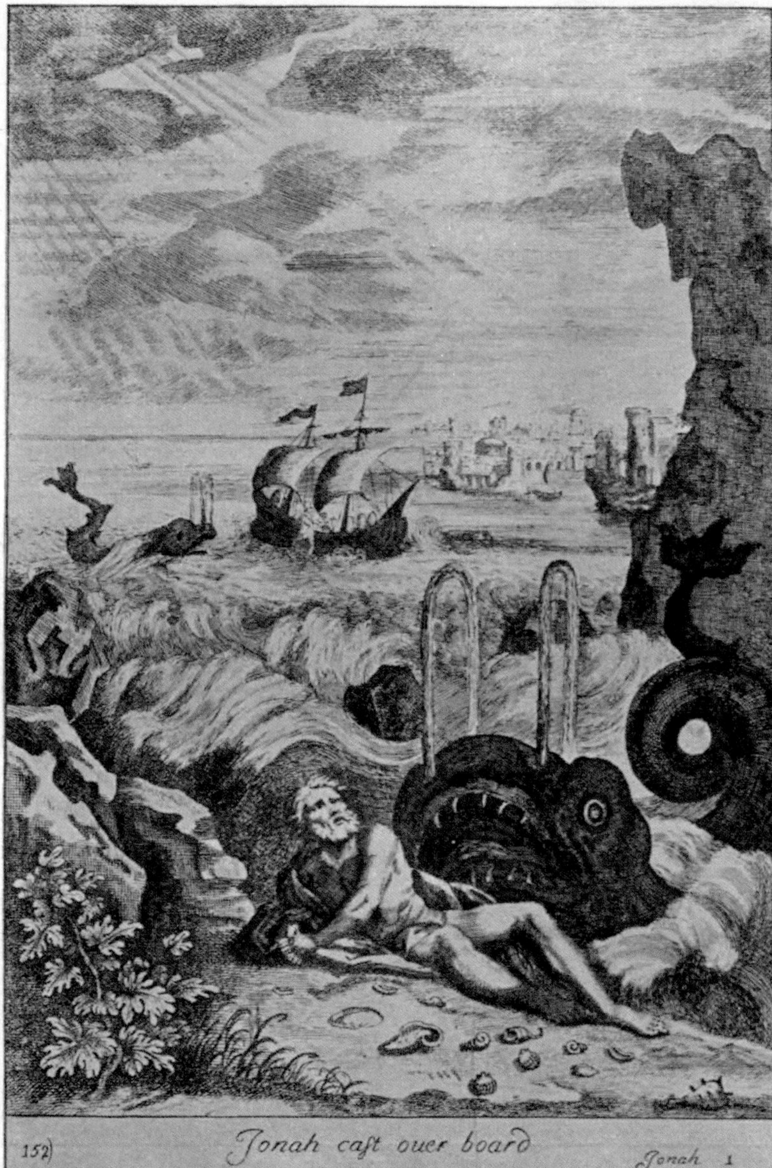

From an old English print.

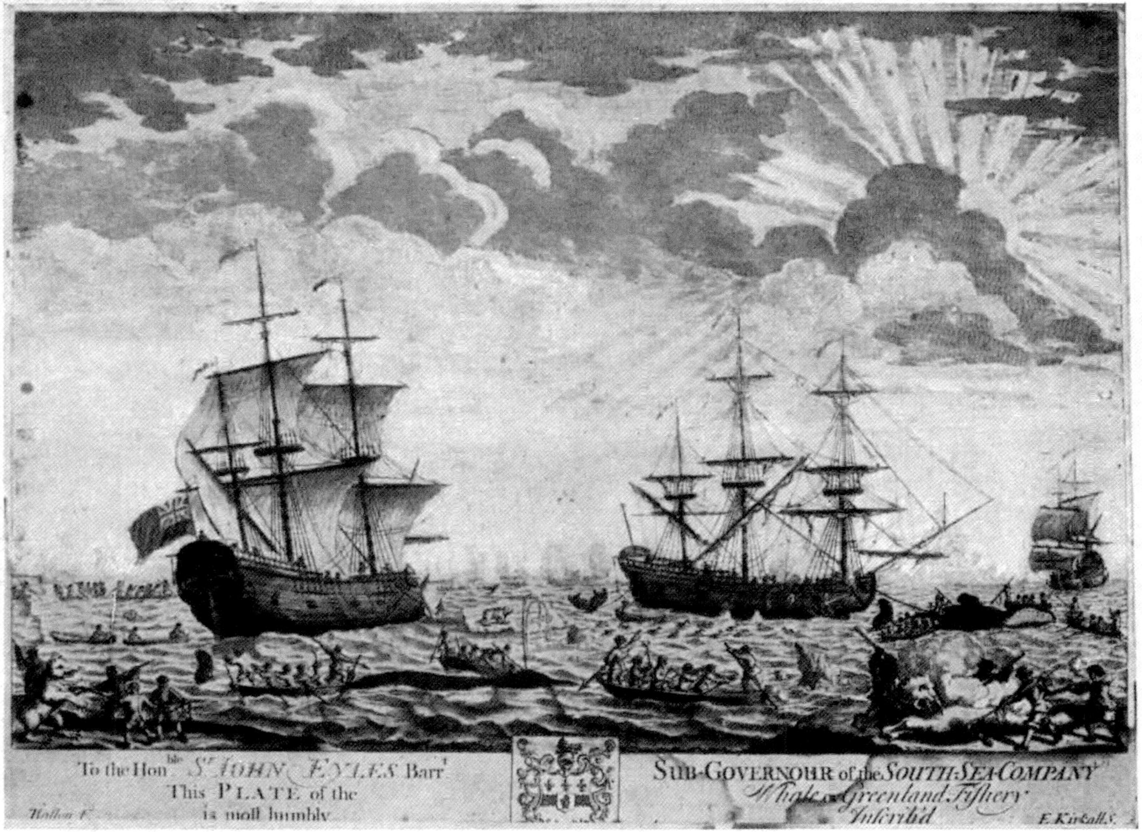

A rare old English print of the Eighteenth Century.

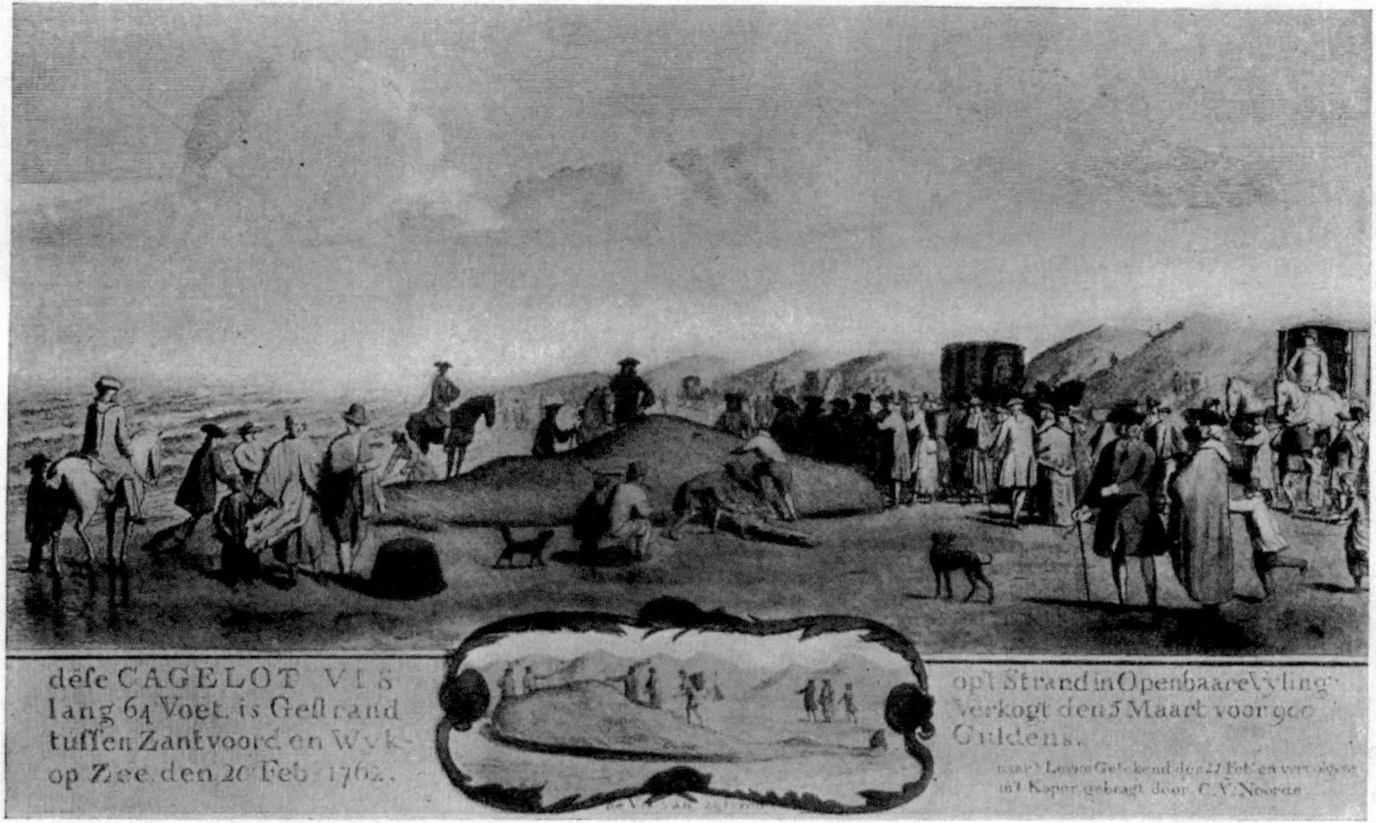

A "cachalot" on the seacoast of Holland. People have always shown intense interest in drift whales.

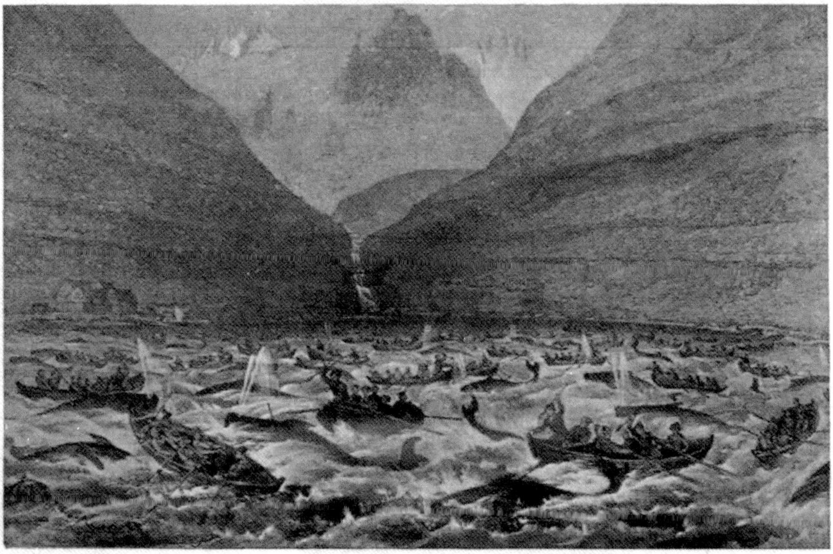

Whale-hunting in Westmannshaven Bay, Norway.
The Norwegians were the earliest whalers of which we have any records.

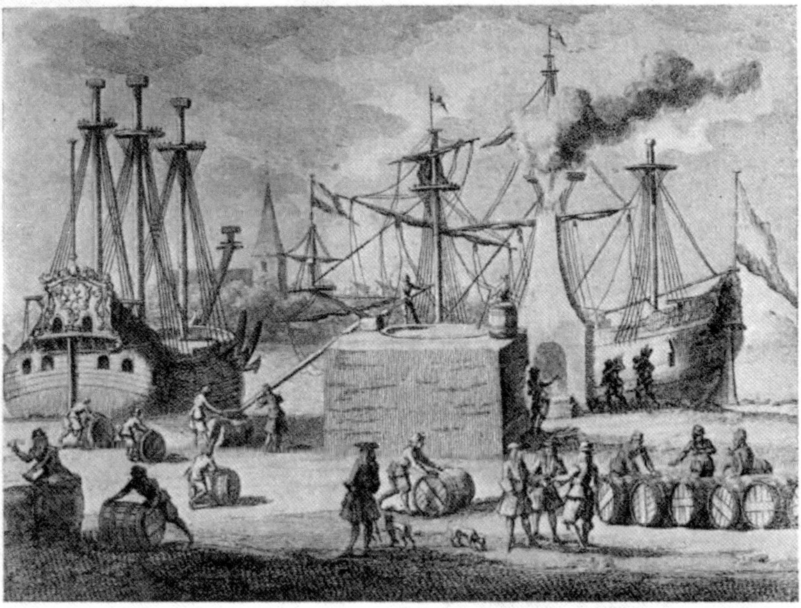

The Dutch boiling oil on shore in a huge "try-works," which was the early method of preparing the oil.

1261, using the whale for food. Also the Icelanders are believed to have whaled some time during the twelfth century. The first reference to English whaling appears during the fourteenth century, and by statutory law the whale was declared "a royal fish." Another curious law was that the King, as Honorary Harpooner, received the head, and the Queen the tail of all whales captured along the English coast, which is very much like halving an apple, there is so little left.

In 1612 the Dutch became the leaders and were still very active about 1680, employing two hundred and sixty ships and fourteen thousand seamen, and during the last part of the seventeenth century they furnished nearly all Europe with oil. To them is attributed the improvements in the harpoon, the line, and the lance, and to their early prominence in the industry we owe the very name "whale," a derivation from the Dutch and German word "wallen," meaning to roll or wallow. They established a whaling settlement at Spitzbergen, only eleven degrees from the North Pole, where they boiled the oil; in fact, during the early days of whaling all nations "tried out" their oil on land. The Dutch continued to be the leaders until about 1770, when the English superseded them owing to the royal bounties.

EARLY NEW ENGLAND WHALING

The history of American whaling really begins with the settlement of the New England Colonies. When the "Mayflower" anchored inside of Cape Cod, the Pilgrims saw whales playing about the ship, and this was their chief reason for settling there. It afterwards proved that the products of the whale formed an important source of income to the settlers on Massachusetts Bay.

The subject of drift, or dead whales which were washed ashore, first attracted the colonists, and there are numerous references to them on record. It was the invariable rule for the government to get one-third, the town one-third, and the owner one-third, and in 1662 it was voted that a portion of every whale should be given to the church. The whale fishery increased steadily, so that in 1664 Secretary Randolph could truthfully write to England, "The new Plymouth colony made great profit by whale killing." The success of the settlers on Cape Cod and elsewhere encouraged Salem to consider ways and means of whaling; for as early as 1688 one James Loper, of Salem, petitioned the Colonial authorities for a patent for making oil, and four years later some Salem whalers complained that Easthamptonites had stolen whales that bore Salem harpoons. As early as 1647 whaling had become a recognized industry in Hartford, Conn., but for some reason did not prosper.

The first white people to explore our New England coasts discovered that the Indians were ahead of them in the pursuit of the whale. The Red Men in canoes attacked these beasts with stone-headed arrows and spears which were attached to short lines. Usually wooden floats were tied to the line, which impeded the progress of the animal, and by

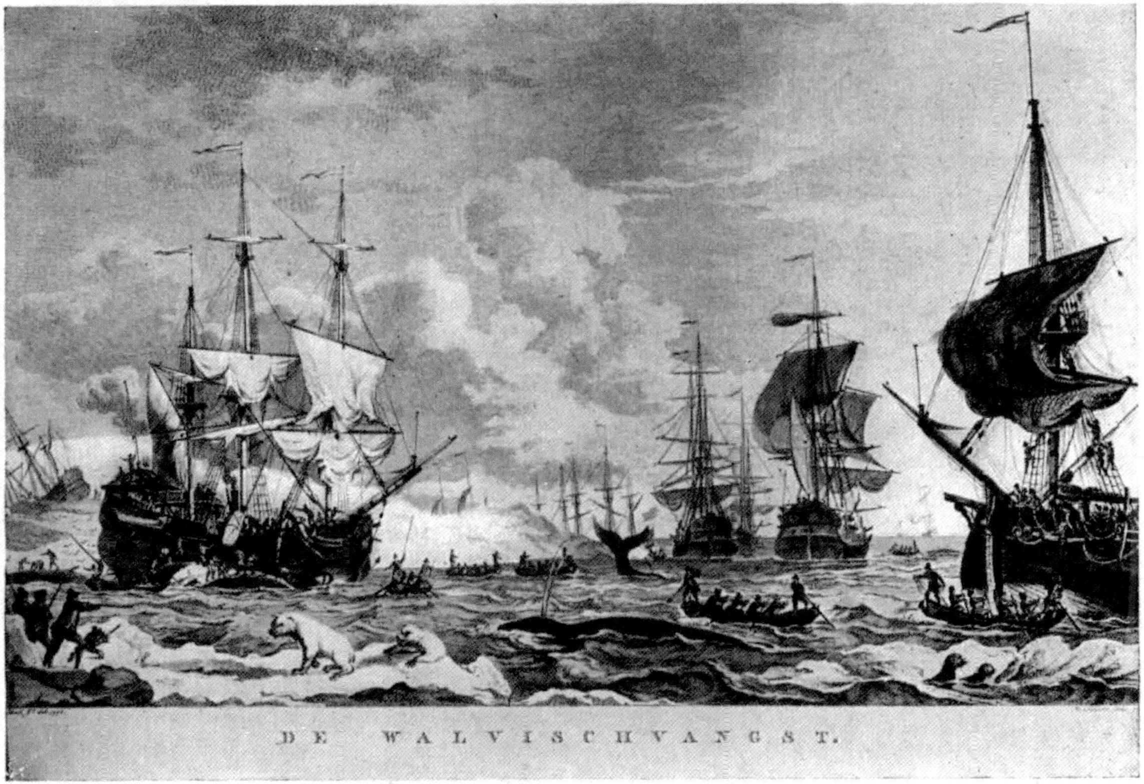

This print shows the high sterns of the old Dutch ships.

frequent thrusts these early hunters actually worried the life out of the whale.

Waymouth's Journal of his voyage to America in 1605 gives the first description of the Indian method of whaling in canoes on the New England coast from November to April, when spouters generally abounded there. "One especial thing is their manner of killing the whale" runs the quaint description "which they call a powdawe; and will describe his form; how he bloweth up the water; and that he is twelve fathoms long: that they go in company of their king

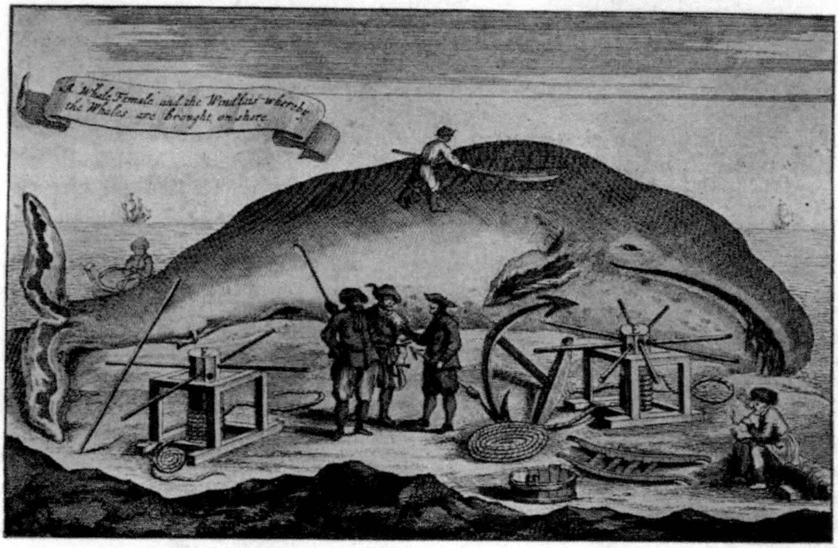

Early method of bringing whales on shore by means of a windlass.

with a multitude of their boats; and strike him with a bone made in fashion of a harping iron fastened to a rope, which they make great and strong of the bark of trees, which they veer out after him; then all their boats come about him as he riseth above water, with their arrows they shoot him to death; when they have killed him and dragged him to shore, they call all their chief lords together, and sing a song of joy; and those chief lords, whom they call sagamores, divide the spoil and give to every man a share, which pieces so distributed, they hang up about their houses for provisions; and when they boil them they blow off the fat and put to their pease, maize and other pulse which they eat."

The Esquimaux at this time were very much more advanced than the Indians, and showed their ingenuity by inventing the "toggle" harpoon, which is in use to this day, and which was improved upon in 1848 by a Negro in New Bedford called Lewis Temple, who made his fortune turning out irons. This harpoon was arranged to sink very easily into the blubber, but when pulled out the end turned at right angles to the shank, thus preventing the harpoon from withdrawing.

Boston is mentioned only occasionally in connection with the Whale Fishery. During 1707 the Boston papers state that a whale forty feet long entered the harbour and was killed near Noddle's Island, and another interesting record is in a letter written in 1724 by the Hon. Paul Dudley, who mentions that he has just received a note from a Mr. Atkins of Boston, who was one of the first to go fishing for sperm whales. There were many whaleships recorded in the Boston records, although fitting out and sailing from other neighboring ports.

NANTUCKET

A large part of the romance of whaling centres around the island of Nantucket and its hardy seamen. It was from here that the Red Men first sallied out in canoes to chase the whale; from here the small sloops first set out laden with cobblestones, as the story goes, to throw at the whales to see if they were near enough to risk a harpoon. These daring Nantucketers were, in 1791, the first to sail to the Pacific, and later on in 1820 to the coast of Japan, and finally they made their ships known in every harbour of the world. Thirty islands and reefs in the Pacific are named after Nantucket captains and merchants.

There is an amusing legend concerning the origin of the island. A giant was said to be in the habit of sleeping on Cape Cod, because its peculiar shape fitted him when he curled himself up. One night he became very restless and thrashed his feet around so much that he got his moccasins filled with sand. In the morning he took off first one moccasin and then the other, flinging their contents across the sea, thus forming the islands of Martha's Vineyard and Nantucket.

From the time of the settlement of the island, the entire population, from the oldest inhabitant down to the youngest child, realized that on the whaling industry depended their livelihood. A story is told of a Nantucket youngster who tied his mother's darning cotton to a fork, and, hurling it at the cat as she tried to escape, yelled "Pay out, mother! Pay out! There she 'sounds' through the window!" The inhabitants always alluded to a train as "tying up," a wagon was called a "side-wheeler," every one you met was addressed as "captain," and a horse was always "tackled" instead of harnessed. The refrain of an old Nantucket song runs as follows:—

"So be cheery, my lads, let your hearts never fail,
While the bold harpooner is striking the whale!"

A young man who had not doubled the cape or harpooned a whale had no chance of winning a Nantucket, New Bedford, or New London belle, and it is stated as a fact that the girls of Nantucket at one time formed a secret society, and one of their pledges was never to marry a man until he had "struck his whale." The well-known Nantucket novel "Miriam Coffin" tells of a girl who made to her two lovers a condition of marriage that they must first of all undertake a whaling voyage, and that she would wed the more successful of the two. It happened that one was a Minister, and the other was no better adapted to the whale fishery; nevertheless, both set out to sea. The former

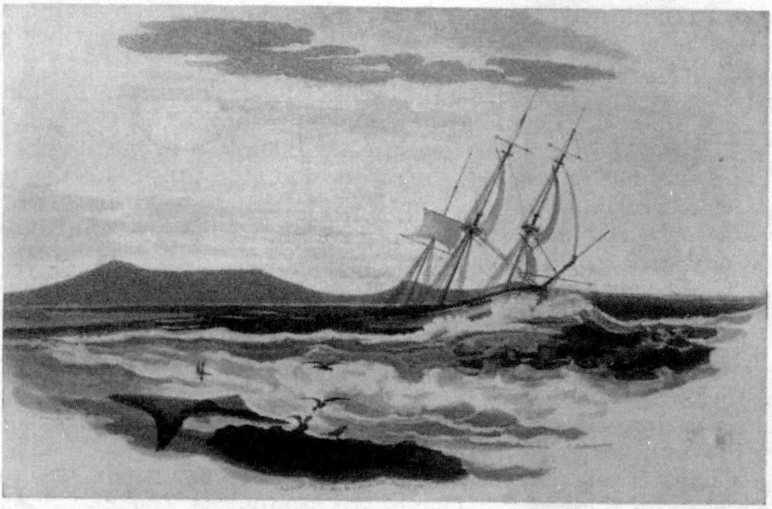

A whaler circling Cape Horn.

was killed by a whale, and the latter returned after an absence of several years, but instead of claiming his bride, he tells her that before going he had already made up his mind that a girl who made such foolish propositions was no girl for him; and so the story ends.

Many a Nantucket bride stepped from her home to her husband's whaleship for a three-year voyage round Cape Horn, which probably suggested these verses:—

"I asked a maiden by my side,
Who sighed and looked at me forlorn,
'Where is your heart?' She quick replied,
'Round Cape Horn.'

. . . .

"I said, 'I'll let your fathers know,'
To boys in mischief on the lawn;
They all replied, 'Then you must go
Round Cape Horn.'

. . . .

"In fact, I asked a little boy
If he could tell where he was born;
He answered, with a mark of joy,
'Round Cape Horn.'"

Any one who did not live in Nantucket was called a foreigner. To show their attitude a schoolboy was asked to write a thesis on Napoleon, and he began by stating that "Napoleon was a great man and a great soldier, but he was an off-islander." In fact, it was an act of condescension for a Nantucketer even to shake hands with a "Mainlander," and there are many of the older islanders to-day who have never set foot on any other soil.

Most of the inhabitants were Quakers, and there was a saying that a Nantucketer was half Quaker and half sailor. Though their cemetery contains about ten thousand graves, there are only half a dozen tombstones in one corner of the field. There are no "Friends" in Nantucket to-day. The following incident shows the Quaker thrift, to which was due in a great measure their success in whaling. When the first chaise was purchased, the owner was about to take a drive in it, but, after a few minutes' deliberation, decided it was too progressive, and would subject him to criticism, so he loaned it only to invalids and funeral parties.

Billy Clark was town crier, and for forty years, up to the time of his death in 1909, he voluntarily announced with a bell and horn the arrival of all whalers and steamers. Once as he went along ringing, a girl asked him rudely where he got his bell, and his reply was, "I got my bell where you got your manners,—at the 'brass foundry.'" Nantucketers declare that his death was due to the fact that he actually "blew his lungs away."

The Chase family has always occupied a most prominent position in the history of the island. One of the family was Reuben Chase, who served under John Paul Jones on the "Ranger," and on his death the following epitaph was placed on his tombstone:—

"Free from the storms and gusts of human life,
Free from its error and its strife,
Here lies Reuben Chase anchored; who stood
The sea of ebbing life and flowing misery.
He was not dandy rigged, his prudent eye
Fore-saw and took a reef at fortune's quickest flow.
He luffed and bore away to please mankind;
Yet duty urged him still to head the wind,
Rumatic gusts at length his masts destroyed,
Yet jury health awhile he yet enjoyed,
Worn out with age and shattered head,
At foot he struck and grounded on his bed.
There careening thus he lay,
His final bilge expecting every day,
Heaven took his ballast from his dreary hold,
And left his body destitute of soul."

Every islander knows the story of the Nantucket skipper who claimed that he could always tell where his ship was by the color and taste of the lead after sounding. Marden, his mate, on one trip determined to fool him, and for this purpose brought some dirt from a neighbor's garden in Nantucket. He woke up the skipper one morning off Cape Horn, and showed him the lead, which had been smeared with this dirt, whereupon, to quote the words of James Thomas Fields,—

> "The skipper stormed and tore his hair,
> Hauled on his boots and roared to Marden:
> 'Nantucket's sunk, and here we are
> Right over old Marm Hackett's garden!'"

Another Nantucket captain always took to sea medicine bottles, each numbered and indexed to suit different complaints. Once his mate was ill, and, looking up the bottle to administer in his case, found that No. 13 contained the cure for his patient. Unfortunately, this bottle had all been used, so, after careful deliberation, he mixed the contents of bottles 6 and 7, which he gave the mate, who promptly died.

Early history tells us that Thomas Macy purchased the island for thirty pounds and two beaver hats, "One for myself and one for my wife," and to him therefore belongs the honor of the settlement of Nantucket; he had been driven away from Massachusetts for sheltering Quakers, which was at that time against the law, and with his friend Edward Starbuck fled to the island and established a colony composed of such well-known families as the Coffins, Husseys, Swaynes, Gardners, Chases, Folgers, and Starbucks. These men were not whalers, but they watched the Indians and learned much from them, and later on employed Ichabod Paddock to come over from Cape Cod and instruct them further.

The character of the island and its situation far out in the ocean, its poor soil, and the number of whales along its shores, all proved an inducement to the Nantucketers to follow the sea as a calling. At first, there were so many whales that they did not find it necessary to go beyond the coast; so, under the guidance of Paddock, lookouts were erected along the South shore, and each man patrolled a certain amount of territory. Each one took his share of whales killed, and business flourished. This method of whaling continued until 1712, when Christopher Hussey, while cruising along the coast, was blown out to sea. He ran across a sperm whale, which he finally killed and brought home. This year was epoch making, as this was the first sperm whale known to have been taken by Americans. The oil from this species of whale being superior to that of all others, the Nantucketers now (1715) decided to change their methods and to whale in the "deep." As the vessels steadily increased in size with greater and greater cargo-carrying capacity, voyages necessarily became longer, extending even to periods of four or five years. In fact, a voyage lasting but two years was considered unusually short. The point of view of most whalers regarding a two-year voyage is shown by the captain who, when boarding his ship, was reminded by a friend that he had not said "Good-by" to his wife,—

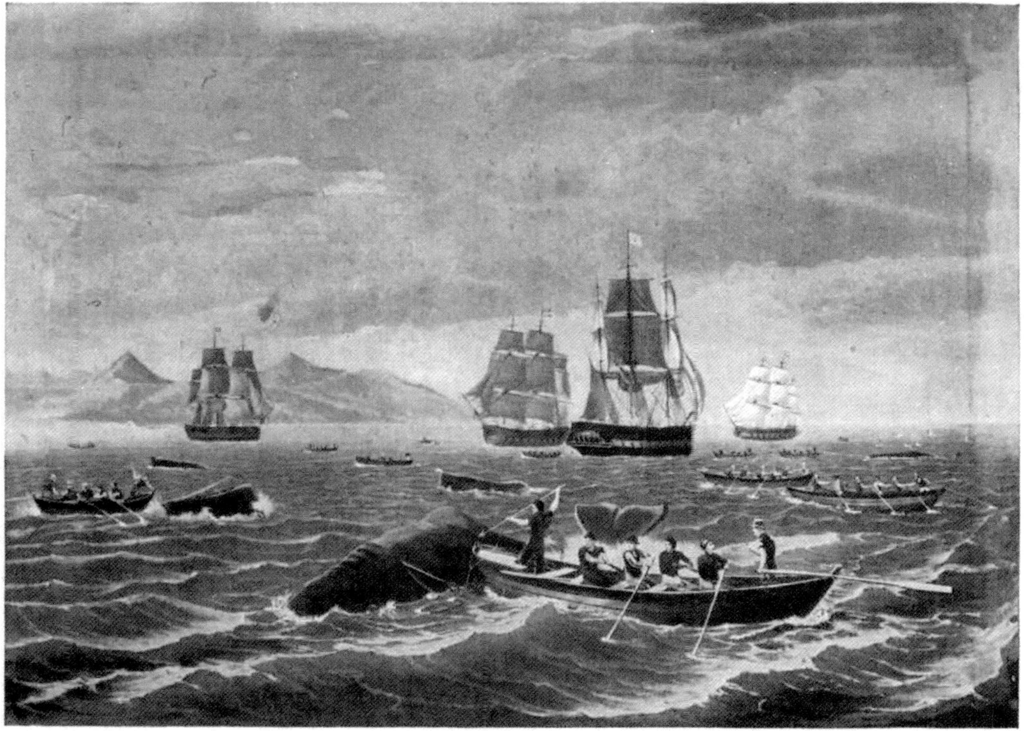

The famous Roach (Rotch) fleet, "Enterprise," "Wm. Roach," "Pocahontas," and "Houqua," among a "school" of sperm whales off the coast of Hawaii. Ships often cruised together and divided the catch. Honolulu owes its rapid rise partially to the frequent visits of the whalers. The first vessel fitted out from the Sandwich Islands was in 1837 and was owned by Henry A. Pierce of New Bedford.

"Why should I?" said he; "I am only to be gone two years."

About 1730 "try-works" were built on the vessels instead of on the shore, and the oil was boiled and stowed away at sea, thus allowing the ships to make much longer voyages. At this time Nantucket owned as many whaleships as all the other ports of America combined. Whaling continued to increase, and the sterile island was turned into a prosperous community, when the Revolution came on, and for the time being practically put an end to the industry. Nantucket was the only port that carried on whaling during the war: the island simply had to whale or starve, as the inhabitants knew no other occupation. Most of their vessels were eventually captured or lost by shipwreck, and over twelve hundred of their men were either killed or made prisoners. The end of the war found the island's business hopelessly wrecked; but, with their usual pluck and determination, the Nantucketers once more built up a profitable fleet. So impoverished were they that the government for one year levied no taxes.

At the close of the war a Quaker, called William Rotch, was Nantucket's greatest whaler, and even he became so discouraged with the prospects at home that in 1785 he left the island in his ship, the "Maria," for London. He endeavored to make some arrangement with the English government to import some whaling families from Nantucket, but, failing to do so, repaired to France, where he succeeded in making an agreement with Louis XVI. A great many families moved to France, and carried on the pursuit from Dunkirk in Normandy. Rotch soon returned to Nantucket, and later moved to New Bedford, where he died. The old Rotch counting-house was later used as a club-room for Nantucket whaling captains, and is even now being used as such. In the old prosperous days this was jocosely called the House of Commons, while another club, which was used by the ship owners, was named the House of Lords.

Immediately after the war, the ship "Bedford," one of the Rotch vessels, was loaded with oil, and sent to England under command of Captain Mooers. This was the first vessel to display the American flag in a British port. It is related that one of the crew of the ship was hunchbacked, and when on shore one day a British sailor clapped his hand on his shoulder, and said, "Hello, Jack, what have you got here?" "Bunker Hill, and be d—d to you," replied the Yankee.

The redoubtable Nantucketers resumed their whaling at the close of the Revolution, and their energy and skill were again yielding rich profits when the War of 1812 almost annihilated the island's fleet. But as it was another case of whale or starve, Nantucket continued to send out a few whalers, and was the only American port during the war that dared to brave the risks of British capture.

About this time, in one of the Pacific ports, an incident occurred which showed in an amusing light the ready wit and intrepid courage of an American whaleman. He had in some way displeased an English naval officer, who, feeling himself highly insulted, promptly challenged the Yankee, who accepted and, being the challenged party, had the

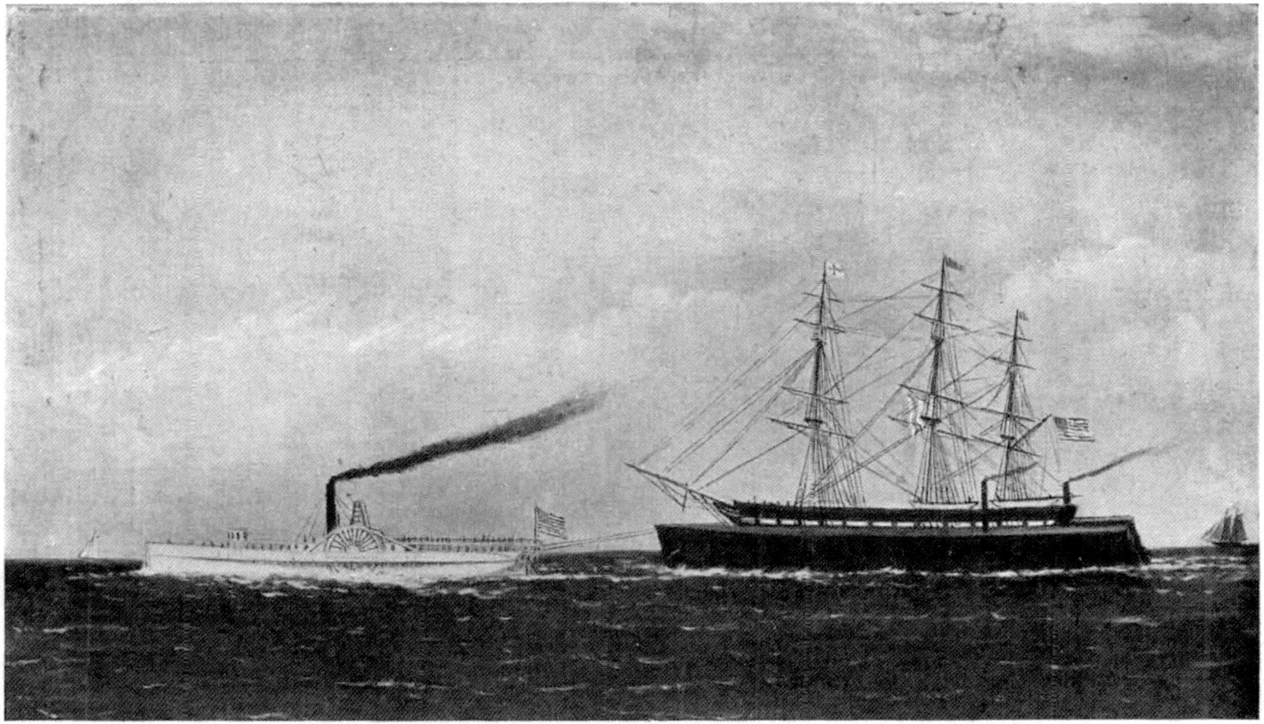

A "camel" floating a whaler to sea over the Nantucket bar. The "camel" was used from 1842 to 1849, enabling the Nantucketers for a time to keep pace with the New Bedfordites.

choice of weapons. He selected, of course, the weapon with which he was most skilful and took his stand with a poised harpoon. It had altogether too dangerous an appearance for the irate Englishman, particularly as the whaleman was evidently an expert in the manual of thrust and parry, and so with as good grace as he could command, the Englishman withdrew from the fight.

At a very early day in the fishery, whaling vessels, which were at first long rowboats and later small sloops, began to increase in size, and about 1820 ships of three hundred tons were found profitable. The increase in profit producing capacity, strange as it may appear, actually sounded the death-knell of the Nantucket whaling, for across the mouth of the harbour ran a bar, over which it soon became impossible for whaling vessels of large size to pass. The difficulty was for a time overcome by the true Yankee ingenuity of some inventive Nantucketer, who devised the "camel," a veritable dry-dock barge in which the larger whaleships, lightened often of oil and bone, were floated over the bar into the forest of masts which in those days characterized a harbour now frequented only by a few schooners and sloops, the small pleasure crafts of the summer residents, and an occasional steamer.

As whaleships still continued to increase in size, the "camel" expedient was only a temporary success; for the time came when vessels were of too great tonnage to be thus floated over the bar, and the daring and skilful Nantucketer, who had taught the civilized world not only how, but where, to whale, had to admit defeat and gradually give up the industry to more fortunately situated ports. At this time, about 1830, Nantucket was commercially the third largest city in Massachusetts, Boston being first and Salem second.

In 1843 Nantucket owned its record number of ships, eighty-eight. In 1846, which is referred to as the "boom" year in American whaling, sixteen vessels cleared from Nantucket and sixty-nine from her near-by rival—New Bedford. In 1869 Nantucket sent her last ship and disappeared from the list of whaling ports. The great fire of 1846 also contributed to the downfall of the industry.

A new era in whaling was to be born, with New Bedford as the centre, and Nantucket was to become only a health resort and mecca for sight-seers, more than ten thousand persons visiting the island in 1914.

NEW BEDFORD

New Bedford undoubtedly owed its whaling success to its proximity to Nantucket, to its wonderful harbour, and to the honesty, thrift, and good business ability of its citizens, most of whom were Quakers.

As in Nantucket, the whole city lived to go whaling, and as each inhabitant made more money, he moved his residence higher up on the Hill. It is said that there was an inn called the "Crossed Harpoons," and another called "Spouter Inn," and there is a Whaleman's Chapel

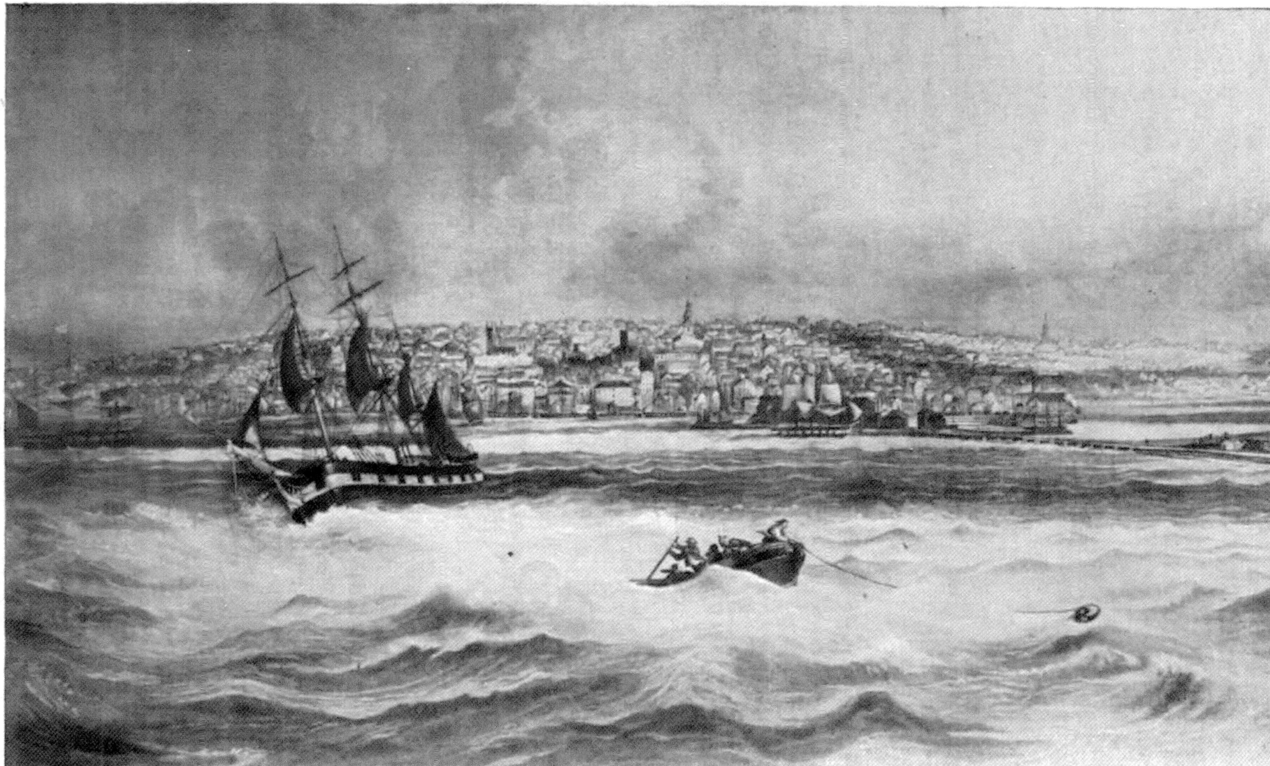

A whaler leaving New Bedford Harbour.

on Johnny Cake Hill where regular Sunday services were held, at which the following hymn was always sung by the congregation:—

"The ribs and terrors of the whale
Arched over me in dismal gloom,
While all God's sun-lit waves rolled by
And left me deepening down to doom.

"I saw the opening maw of hell,
With endless pains and sorrows there;
Which none but they that feel can tell—
Oh, I was plunging to despair—

"In black distress I called to God,
When I could scarce believe him mine,
He bowed his ear to my complaints—
No more the whale did me confine."

The pulpit of this chapel was made to represent the prow of a whaleship, and was ascended by means of a rope ladder, which the minister, who had been a harpooner in his youth, hauled up after him. Around the walls of this little church can still be seen tablets erected in memory of many whalemen who lost their lives at sea. There also was a daily paper called *The Whaleman*, which gave the reports of the whaleships and the whaling news. It has been said that New Bedford fathers gave whales for dowers to their daughters, and that they had reservoirs of oil in their attics to burn on gala occasions.

It is a curious fact that three Morgans not long ago married three Rotchs, three Rotchs married three Rodmans, and three Rodmans married three Motleys. Among other well-known New Bedford whaling families are the Hathaways, Swifts, Howlands, Morgans, Stones, Delanos, Rodmans, Seaburys, Giffords, Tabers, Grinnells, and Wings.

Whaling was a tremendous financial gamble, and until a vessel came home "clean" or "greasy," meaning empty or full, the success of the voyage was not known. They tell a story of a New Bedford captain who had been out for nearly four years, and as he came up to the wharf the owners asked him what luck he had had. His reply was, "I didn't get any whales, but I had a damn good sail." There is another tale of a seaman whose vessel left New Bedford on the day of his mother's funeral. Naturally he set sail with a heavy heart, and during his three years' cruise he thought many times of his sorrowful father at home. As the ship neared the docks he was met by his father with "Hurry up, Jim, I want to introduce you to your new mother." There were many changes at home during a long cruise, and sometimes even the fashions had entirely changed. One whaleship captain described his surprise at seeing for the first time the crinoline or hoop skirt.

The real founder of New Bedford, and the pioneer of the whale fishery at this port, was Joseph Russell, who sent his ships out in 1765. Several years later the first ship was launched and was called the "Dartmouth," and this vessel is well known to history owing to the fact that she was one of the ships that carried into Boston Harbour the tea that was thrown overboard. The whaling industry increased steadily, except during the wars, until 1857, when the New Bedford fleet numbered three hundred and twenty-nine vessels, was valued at over twelve million dollars, and employed over twelve thousand seamen. If these vessels had been strung out in line, they would have stretched over ten miles. In addition to these sailors, thousands of others were employed at home making casks, irons, ropes, and many other articles used in whaling. In fact, it was often stated that the population was divided into three parts,—those away on a voyage, those returning, and those getting ready for the next trip.

There were many nationalities represented in the crews of the whalers, and the New Bedford streets presented a very foreign appearance, with Spaniards, Portuguese, Dutch, Norwegians, Germans, French, English, Scotch, Irish, Sandwich Islanders, and New Englanders at every turn. A large number of Portuguese served on whaleships, and a part of the city near the south end of Water Street became known as Fayal.

The "Golden Age," as it is called, of whaling was between 1825 and 1860, and during the whole of this period New Bedford assumed the lead, even long after other ports had given up the pursuit. It is estimated that about the year 1848 there were over seventy millions invested in the industry and seventy thousand persons derived from it their subsistence.

It is an interesting fact that the insurance on American whalemen was about one-half the rate that was charged the Englishman, which certainly showed the superiority of our Yankee seamen. There were several whaling insurance companies in New Bedford. There is a story told of a New Bedford ship owner who had just heard that his vessel had gone down and he hadn't yet received the insurance policy from the company. He sent a letter down to the office which read as follows: "I have heard from my ship and thee need not place the insurance." Of course, the policy was sent up immediately.

The New Bedford whalers explored new grounds, and to this fact chiefly is due the continued prosperity of its whale fishery, but it was to die slowly; in 1875 the fleet from this port had declined to 116 vessels, in 1886 to 77 ships, and in 1906 to 24.

One of the chief historical events of New Bedford happened in 1861, when the famous Stone Fleet sailed from that port. The United States government decided to purchase some old ships and sink them in the channels of the harbours of Charleston and Savannah, to prevent blockade running during the war. H. Bartlett & Sons supervised their purchase and Captain Rodolphus N. Swift was the general agent. Bartlett purchased some of the old whalers for as small a price as thirty-one hundred and fifty dollars, some of them having more cement than wood in their hulls. To James Duddy, a teamster, fell the task of supplying the seventy-five hundred tons of stones with which to fill the vessels, and many a New Bedford stone wall now lies at the bottom of some of our Southern harbours. Captain Rodney French, an old

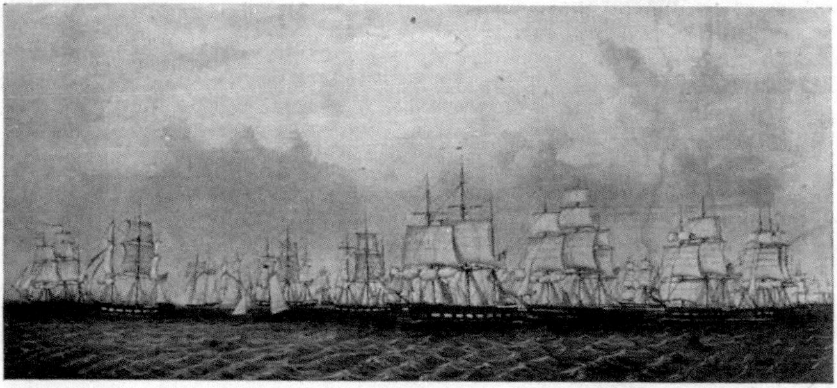

The famous Stone Fleet sailing from New Bedford, Nov. 16, 1861. The ships were loaded with stones and were sunk in the mouths of certain Southern harbours during the Civil War, to prevent blockade runners from entering. The vessels in this picture are the Garland, Maria Theresa, Rebecca Simms, Leonidas, South America, Archer, American, Harvest, Amazon, Cossack, Courier, Henrietta, Potomac, Kensington, Herald and L. C. Richmond.

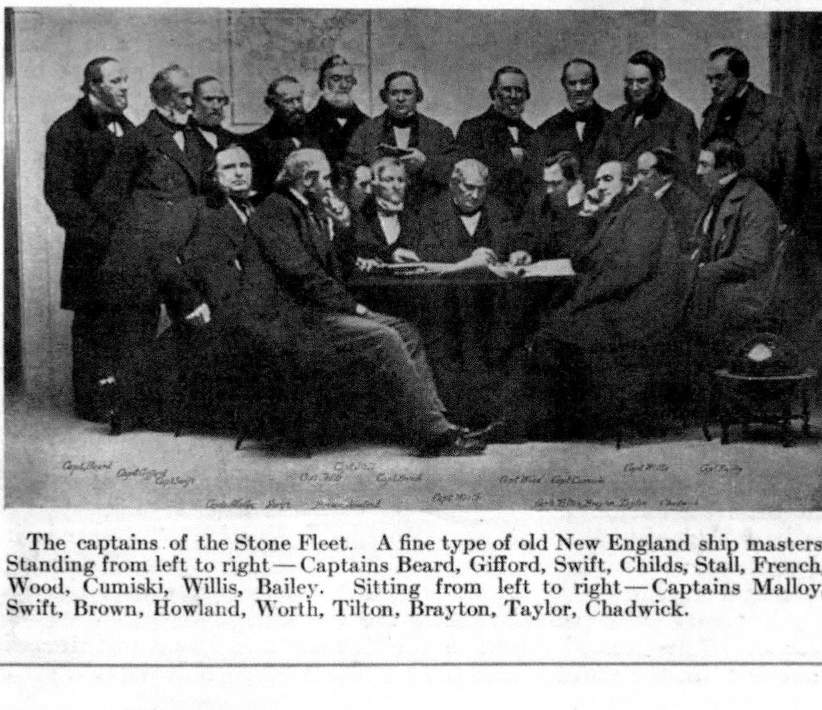

The captains of the Stone Fleet. A fine type of old New England ship masters. Standing from left to right—Captains Beard, Gifford, Swift, Childs, Stall, French, Wood, Cumiski, Willis, Bailey. Sitting from left to right—Captains Malloy, Swift, Brown, Howland, Worth, Tilton, Brayton, Taylor, Chadwick.

"slaver," who afterwards became Mayor of New Bedford, was selected as commander of the fleet, and on Thanksgiving Day most of New Bedford assembled on the wharves and saw fifteen of her once famous fleet, which had for years been the homes of its seamen, sail forth never to return. It must have been a very sad day for the city, and it may be said that this event marked the beginning of the decline of the industry at New Bedford.

One captain insisted upon washing the decks of his ship every morning, using pulverized stones instead of sand, and another, to give the fleet a warlike appearance, mounted a formidable "Quaker" gun, made from a section of a spar.

A second fleet sailed later in the year, making forty-five vessels in all. Although the expedition cost the government about a quarter of a million dollars, its success was only temporary. The captain of the "Alabama" swore vengeance on New Bedford and destroyed or captured every whaler he could find, and in the "Alabama" awards that were made after the war New Bedford received a large share.

Ten years later occurred the worst disaster that ever befell a whaling fleet. Thirty-four whalers were caught in the ice in the Arctic regions and sunk, and it is a curious fact that, while the loss reached one million five hundred thousand dollars, not a single human life was sacrificed. These three pictures of a series of five on the following page show the sinking of the ships, the abandonment of their vessels, which had their flags union down, and the eighty-mile sail through the ice-floes to the open sea, where twelve hundred and nineteen men, women, and children were taken home in the seven whalers that had not been lost in the ice. It must have been very crowded, as each ship had to stow away several hundred persons in addition to her own crew. There were many sad hearts as they left their vessels and almost all of their belongings, and started off in the small boats. The trip to sea and the transshipment in the heavy swell must have been made with the utmost care, otherwise many lives would have been sacrificed. The loss to the New Bedford owners was so tremendous that they never really recovered from the catastrophe, and many families had to economize for years after. The Swifts, Howlands, and Rotchs were among those who lost ships.

On one of the vessels in the first picture of this series was a large quantity of the finest Manila cigars and also some rare Madeira wine, that had been picked up in the Philippines the year before on instructions from the ship's owner. When the captain of this vessel reached New Bedford and reported the loss of his command, the owner's first question, after listening to the dismal tale, was whether his cigars and wine had been saved. "All of it," came the reply. "Where is it?" said the owner, looking more cheerful. "Well, you see, I drank the wine and Mr. Jones, the mate, he smoked the cigars, and they certainly done us both good," replied the captain.

The ship "Progress," shown in the last picture, forms an interesting connecting link between the Stone Fleet and this 1871 disaster.

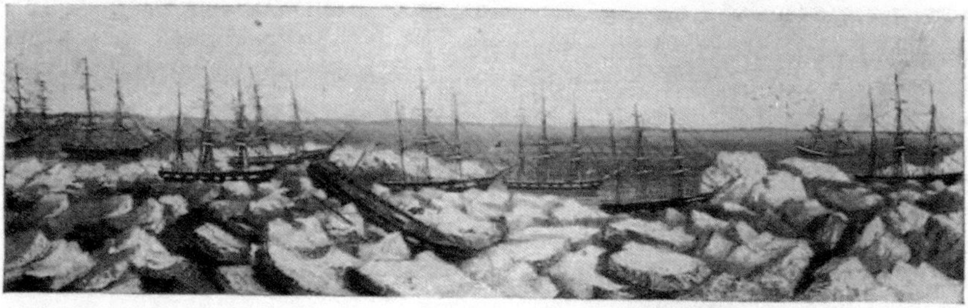

Abandonment of the whalers in the Arctic Ocean, September, 1871. Vessels surrounded by the ice, and many of them in a sinking condition.

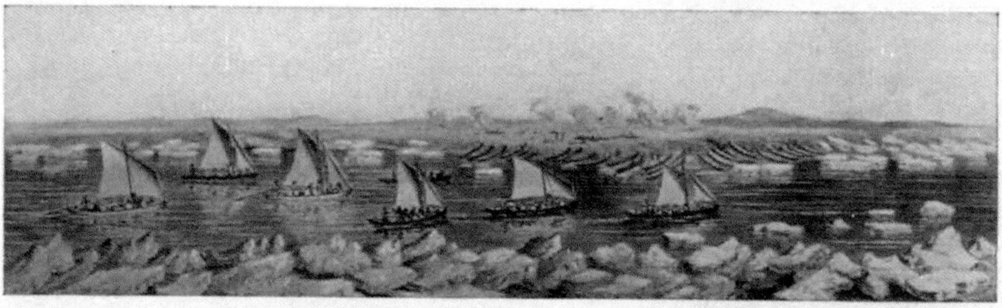

Abandonment of the whalers in the Arctic Ocean, September, 1871. Showing the whaleboats being hauled up on Blossom Shoals, where the ship-wrecked crews spent the night crowded under the upturned boats.

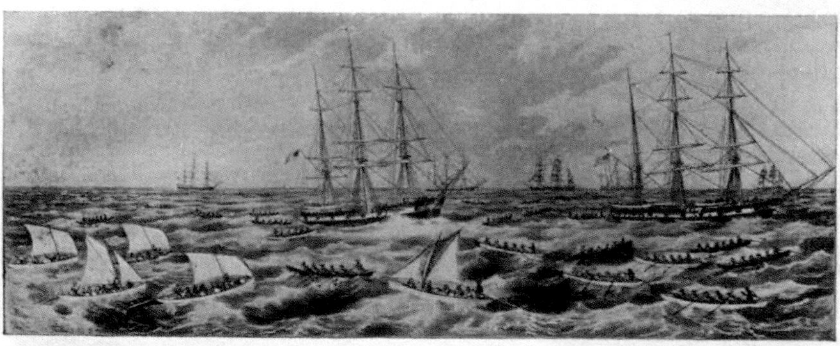

Abandonment of the whalers in the Arctic Ocean, September, 1871. The seven ships receiving the 1217 men, women, and children of the abandoned vessels. The sea was very rough and the trans-shipment was very dangerous. The ship "Progress," whose history is given on the opposite page, is at the right of the picture.

Under the name of the "Charles Phelps" she whaled from Stonington, Conn., for a number of years and finally was purchased for the Stone Fleet. She was found to be in such good condition that the government decided not to sink her, and she returned to New Bedford and was sold; and it was this same vessel that took part in the rescue of the twelve hundred and nineteen shipwrecked people ten years later. In 1893 she was fitted out as if for a whaling voyage and towed by way of the St. Lawrence River to Chicago, where she was exhibited at the Fair, and now lies rotting on the sands of the lake at South Chicago. No other whaler ever had so interesting and varied a history.

The year after this Arctic disaster found the fleet again in the Arctic, and the "Minerva," one of the ships left at Point Belcher, was discovered and found to be in good condition; the others had sunk. One lone person was found who had remained on board his ship for the whole year, and his sufferings had been fearful. The natives had stolen all the whalebone and oil from the sinking vessels, and when some of the same shipwrecked captains arrived the next year the Esquimaux tried to sell them back their own property, and one native was using one of the chronometer cases as a dinner pot in which to boil his blubber. The "Minerva" was manned and sailed to New Bedford and continued in the whaling industry.

New Bedford ships suffered severely during the Rebellion, but later new ones were added to the fleet and business again prospered. Lack of space prevents enumerating the achievements of American whalers during the Civil War. Captain William P. Randall, however, will go down in history as a hero of this war; he was brought up on a whaleship and later served in the navy.

Captain Frederick Fish, father of Frederick P. Fish and Charles H. Fish, of Boston, was one of the best known and most respected of the whaling captains sailing out of New Bedford. He commanded the "Montreal" and the "Columbus" when only twenty-two years old, made nine voyages round the world, and was one of the most successful whalers of his day. Once when near the Sandwich Islands his vessel happened to anchor very close to an English ship, and Captain Fish noticed that every evening at sunset the English commander, while at anchor, set all sails and then furled them again in order to show how quickly this work could be performed. After a few evenings Captain Fish ordered his crew to do the same, and the time consumed was so much less that the next evening the Englishman decided he did not care to go through the performance; in fact, he never tried to show off again in that port.

There is also another amusing story told about Captain Fish. His ship at one port took on a great many chickens, which were used for food, and finally one of the crew rebelled and informed the captain that he had eaten enough hen. He was immediately ordered out on a yardarm and was made to crow like a rooster for such a long time that when he was again allowed on deck, he had a most excellent appetite for another chicken dinner. Captain Fish delighted in telling of the time when he took a local pilot on board somewhere in the Pacific to conduct his vessel into port. He asked the navigator if he were sure of his course, and received a prompt and decisive answer in the affirmative. Presently, to the disgust of the captain, the vessel touched. The next question put to the pilot was whether or not he could swim, and finding that he could, Captain Fish ordered his crew to throw him overboard. This was done, and, the distance being short, the swimmer made the land, and the captain himself took his vessel in the rest of the way.

Captain Fish was an excellent story teller, and another yarn has been handed down in connection with one of his trips. The voyage had been

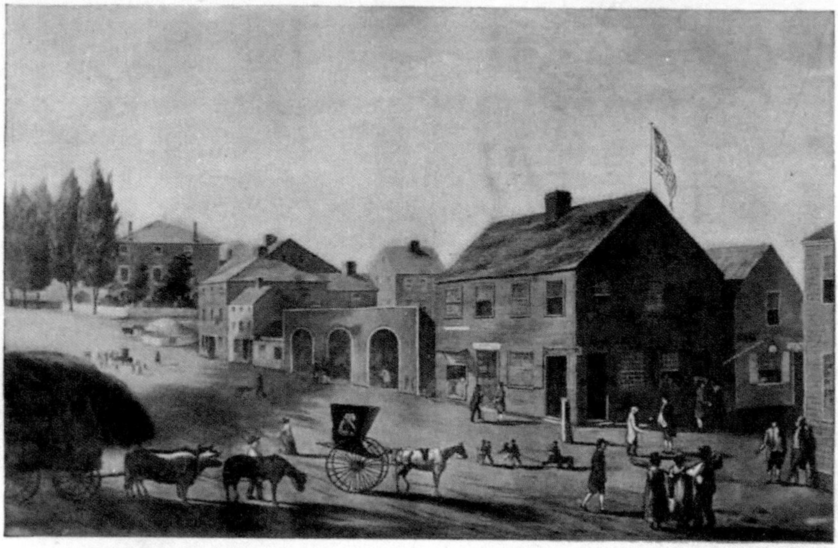

New Bedford fifty years ago (1808). (This print is dated 1858.)

very unsuccessful, and as he was looking over his chart he tossed his dividers down in a disgruntled manner, and by accident they chanced to stick in the chart. He then conceived the novel idea of sailing to the very place where his instrument happened to land, and curiously enough he was rewarded by a very large catch.

Once when one of his whaleboats had been overturned by a fighting whale he hurried to the assistance of the crew, who were struggling in the water, and to his amazement found two of them squabbling over the ownership of a pair of old shoes, instead of thinking about saving their lives. It is a curious fact that he never learned to swim, and often saved his life when capsized by grabbing some floating débris. His nerve and courage were remarkable, and it is related that even on his death-bed he told the doctor an amusing story.

This picture of New Bedford in 1808 is most interesting. The oil market shed on the right-hand side of the street was built in 1795 by

Barnabas Russell for his son Joseph, and the last building shown on the right of the picture was the mansion of William Rotch, Sr., and the first estate in the village at that time. This Rotch was the son of Joseph Rotch, one of New Bedford's earliest whalers, and he himself is represented in his old chaise, the only private carriage then in the town. He is negotiating for a load of hay, and from all accounts he must

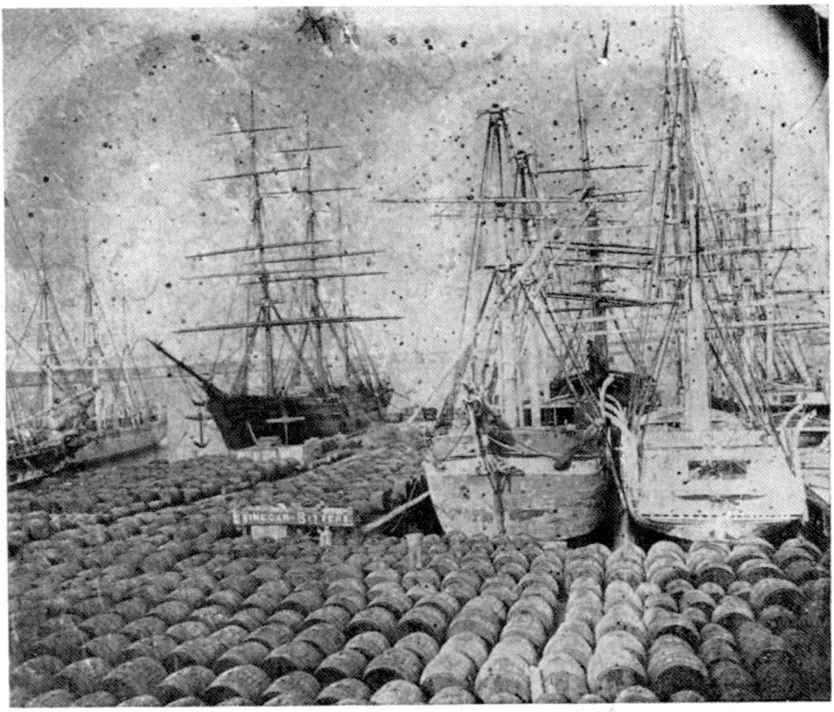

Oil stored on the wharves at New Bedford awaiting a favorable market. The owners, dressed in silk hats, long-tailed coats, and polished top boots, might often be seen watching, testing, and marking the oil-barrels.

have been a keen business man, for he was often seen going to market so early that he had to use a lantern. All the other figures in this picture also are intended to represent well-known citizens of the time. The two men shaking hands are Captain Crocker and Samuel Rodman; the latter, who was the son-in-law of William Rotch, had the reputation of being the best dressed man in New Bedford in his day. One of the boys harnessed to the small cart is the Hon. George Howland, Jr., great-uncle of Llewellyn Howland. H. H. Hathaway, Jr., and Thomas S. Hathaway have three ancestors in the picture.

OTHER NEW ENGLAND WHALING PORTS

Rhode Island pursued whales in 1731, Newport and Providence being the two most successful ports. Fifty ships were owned by Connecticut and Rhode Island in 1775. Massachusetts owned over three hundred at this time. Rhode Island was more of a "slave" than a whaling State. New London became a great whaling port in 1846, and was the third in importance in New England.

The people of Cape Cod began sending ships to sea about 1726, and a few years later a dozen or so vessels were fitted out at Provincetown. Boston claimed twenty whaleships in 1775, and registered from one to eleven vessels almost every year until 1903, since which date no whaleship has been recorded from this port. Gloucester turned to whaling in 1833.

The following figures show the different whaling ports in Massachusetts and the largest number of vessels enrolled in any one year in each. New Bedford, of course, held first place with 329 in 1857, with Nantucket 88 in 1843. Provincetown claimed 54 in 1869; Fairhaven 50 in 1848 to 1852; Edgartown and Mattapoisett owned 19 each; Salem had 14 in 1840; Boston 11 in 1868; Dartmouth, 10; Plymouth, 9; Falmouth, 8; Wareham, Fall River, and Marion, 7 each; Beverly, Holmes' Hole, Orleans, 5 each; Lynn, 4; Newburyport, 3; Gloucester, Dorchester, and Sandwich, 2 each; and the following claimed 1: Braintree, Hingham, Marblehead, Barnstable, Duxbury, Quincy, Truro, Yarmouth, and Wellfleet. Of the Rhode Island towns Warren owned 25; Newport, 12; Bristol, 10; Providence, 9. Connecticut towns that owned whalers were New London, 70; Stonington, 27; Mystic, 18; and a few scattered among half a dozen other places. Portsmouth, N.H., at one time owned two vessels, and between the years 1835 to 1845 Bath, Bucksport, Portland, and Wiscasset in Maine each had one. Massachusetts, however, could claim five-sixths of the total fleet.

A few words must be said in praise of Samuel Mulford of Long Island. Governor Hunter of New York claimed for his State a share of all whales caught, whereupon Mulford waged war against this act in every possible way. Finally he sailed to London and put his case before the Crown. The people in London were much amused at his country clothes, and the pickpockets in particular became a nuisance to him in the streets. Mulford, however, showed his resourcefulness by sewing fish hooks in his pockets and succeeded in capturing the thief. Another incident shows the ingenuity of the whaleman. The ship "Syren" was attacked by a horde of murderous savages, and the crew of the ship would, doubtless, have been murdered had it not been for a quick stratagem of the mate. He remembered a package of tacks in the cabin and yelled, "Break out the carpet tacks and sow 'em over the deck." The natives, yelling with pain, jumped headlong into the sea, and the ship was saved.

The world owes many discoveries to the energy and determination

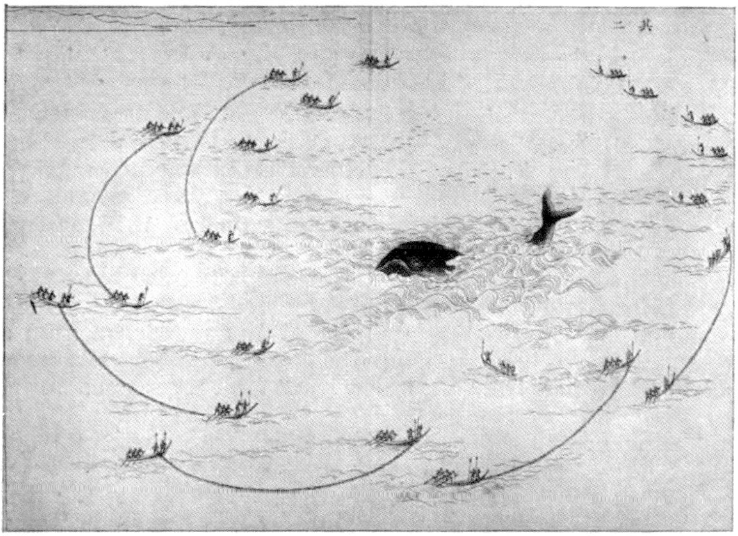

The Japanese method of capturing whales was to entangle them in nets. A great many boatloads of men would drive the whale toward the nets by throwing bricks and stones at it. When once entangled the infuriated animal could be easily killed. In 1884 the Ukitsu Whaling Company employed over 100,000 whalemen. One of the most successful of the Japanese in this pursuit was Masutomi Matazaemon, who accumulated a large fortune. The Japanese have been very slow to adopt our Western methods.

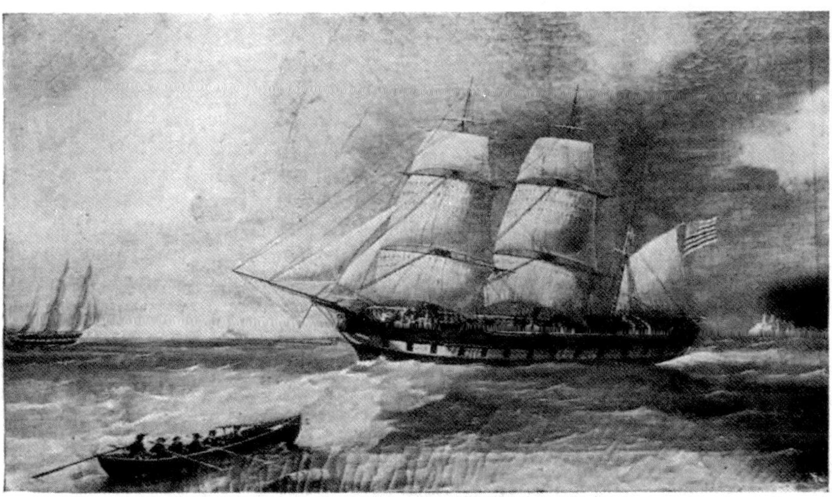

A typical "blubber hunter" cruising for "right" whales in the Arctic.

of whaleship captains. Over four hundred islands in the Pacific were discovered and named by American whalemen, and the history of New Zealand is closely connected with the visits of New England whalers. Australia, too, was opened to the world by the whalemen.

It was to a certain extent due to the testimony of Captain Bryant, a whale captain of Mattapoisett, that Alaska was purchased by the United States government. That there was a northwest passage was also discovered by American whalemen in this way: the date and name of a ship were always marked on its harpoons, and in several instances whales were captured in the Pacific by ships that were known to have been cruising not long before in the Atlantic. It was Captain Timothy Folger, of Nantucket, who charted the Gulf Stream at the request of Benjamin Franklin, to whom he was related, and this drawing was engraved on an old chart and preserved in London. In this way English mariners discovered how to avoid the swift current and thereby gain much time. Our seamen in the early days were not very kindly treated by the Japanese, but, finally, several whalemen secured their good will by teaching them English. This encouraged the American government to send out Commodore Perry's expedition, which succeeded in making our first treaty with Japan, thus opening that country to Western civilization.

It was difficult to make discoveries ahead of our whalemen. In 1834 two Russian discovery ships approached a forlorn little island in the Antarctic Ocean and the commander was about to take possession in the name of his Czar. There was a dense fog at the time, but when it cleared away they were very much surprised and vexed to see a little Connecticut ship at anchor between their two vessels. The name of this whaler was the "Hero" of Stonington, captained by Nathaniel B. Palmer, who was only twenty-one years of age and was just returning from his discovery of the Antarctic Continent. The Russian commander was so impressed by the achievement of this youthful captain that he cheerfully acquiesced in naming the place Palmer's Land. This name has since been changed to Graham Land. It is an undisputed fact that the whalers prepared the way for the missionaries.

ABOARD A "BLUBBER HUNTER"

Nothing can be more romantic than to be attending a clam-bake on Mishaum Point or Barney's Joy and to see a whaleship, or "blubber hunter" as she is often termed, round the point and start to sea. It is with quite different feelings that one peers down into her forecastle, which is often referred to as the Black Hole of Calcutta. This room, which is the home of thirty to forty men for three or four years, is reached by a perpendicular ladder through a small hatchway, which is the only means of ventilation. The bunks are in tiers and are about the size of a coffin, so narrow that it has often been said that one has to get out of them in order to turn over. A small table in the centre of this "hole" and the seamen's chests lashed to the floor comprise all the furnishings, except possibly a few bottles of rum, which were often labelled "camphor." In fact, one might speak of the dis-accommodations of the forecastle, and it is no wonder that a cruise in a whaler is often spoken of as a "sailor's horror." The odor of grease, dirt, oil, and lack of air are unbearable except to one thoroughly accustomed to a whaling trip, and sailors often say that this attractive place should not be approached without a clothespin on one's nose. The utensils comprised a few tin plates and a bucket of water, with one cup for the use of every one. The food consisted of "longlick" and "scouse," the former made of tea, coffee, and molasses, and the latter of hardtack, beans, and meat. It is not difficult to see, therefore, why most of the captains anchored their ships well out beyond the harbour, so as to prevent desertions after the novice seaman had glanced at his sleeping quarters. There have been cases of sailors jumping overboard on the chance of reaching land, and it is on record that the greater part of a whaleship's crew once floated to shore on the cover of the try-works. A captain was very careful where he allowed his men to land, and, in case he was afraid of desertions, took care to allow them shore leave only at places where the natives were troublesome, or where for a ten-dollar bill he knew he could get the whole crew returned to him.

The whaleship looked very clumsy and was built for strength rather than for speed, the bow and stern looking as if they were made by the mile and chopped off in lengths to suit. It is a curious fact that the "Rousseau," belonging to the Howlands, when caught in a storm off the Cape of Good Hope sailed astern for seven days faster than she had ever sailed ahead, and successfully weathered the point.

There is an amusing anecdote that has gone the whaling rounds, of a greenhorn, called Hezekiah Ellsprett, who arrived on board the night before sailing. One of the men told him that the first ones on board had the right to pick out their berths and suggested that he paint his name on the berth he should select. Hezekiah looked round, found the best-looking cabin, painted his name in big letters on the outside of the door, and made himself comfortable for the night. He had chosen the captain's room, and in the morning the captain came on board, and in very violent terms informed him that he was in the wrong end of the ship.

The whaleman's life was indeed a hard one, and his share of the profit, or "lay" as it was called, was so small that at the end of a moderately successful voyage if his share amounted to several hundred dollars he was doing well. His earnings were depleted by the captain's "slop chest," where the sailors had to purchase their tobacco and clothes at high prices, and if there were any kicks the answer was that he could "get skinned or go naked." The most necessary part of the sailor's equipment was the sheath knife which was used about the ship and to repair his clothes, and it was this same implement that he used to cut his food!

Regular deck watches were kept, and in good weather the officers often winked their eyes if some of the men slept. Among sailors this was called a "caulk," and often some kind of a joke was played on the sleeper. In one case they tied a live pig to the slumberer's feet and watched the fun from behind the try-works.

Whalers would rarely cruise past the Azores without stopping at Fayal, where they were most hospitably received by the American Consul, who for centuries was one of the Dabney family. In fact, the island is often referred to among whalemen as the "Isle de Dabney."

"Gamming" or exchanging visits between two whalers at sea was thoroughly enjoyed and gave a chance to the sailors to swap experiences, and many a weird, sorrowful, or wonderful story must have been related. An incident is recorded of a meeting between two brothers who had lived in Nantucket, and who had not met for twenty-three years. There is an old adage among whalers that when a year from home, on "gamming" with a ship that has sailed subsequent to your own departure, you have the privilege of begging; when two years out, of stealing; and when three years away from home, of both stealing and begging.

A New London ship was once holding a reception on board for some natives, and each of the crew was endeavoring in some way to amuse the guests. One seaman took out his set of false teeth, thinking he would provide entertainment; but instead the natives became so alarmed that they tumbled over the side into their canoes and made their retreat as quickly as possible. The crew was asked on shore for a return visit; but an invitation to the exhibitor of the teeth was not forthcoming, and he was obliged to remain alone on the ship, much to his disappointment. Captain Gardner of Nantucket stated that in thirty-seven years he spent only four years and eight months at home, and Captain North, also of Nantucket, figured that he had sailed one million one hundred and ninety-one thousand miles.

Nothing could have equalled the joy of returning home after a long voyage, and the anxiety to reach port was almost unbearable. Often a vessel ran into bad winds and had to anchor for days a few miles off shore, and there is one case known of a ship being blown to sea and lost after having actually come within sight of New Bedford Harbour.

Many a whaleman has laughed at this story. It was customary for the first mate to keep the log book. One day he was intoxicated, so the captain entered the day's events, noting that "the mate was drunk all day." The next day the mate protested, but the captain said that it was true and must remain on the records. The mate resumed his charge of the diary, and got more than even with his superior officer by recording on the following day that "the Captain was sober all day."

WHALING IMPLEMENTS AND WHALEBOATS

The earliest method of killing whales was by means of the bow and arrow, and the first accounts of New England whaling refer to the harpoons as being made of stone or bone. There are three kinds, however, that have been popular among American whalemen: one had one barb

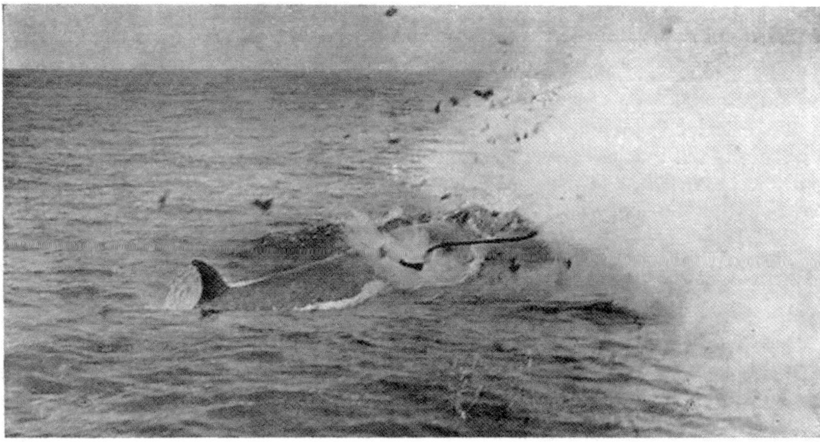

This picture, taken by Roy C. Andrews, Esq., of the American Museum of Natural History, on his last whaling expedition, shows a bomb exploding in a whale.

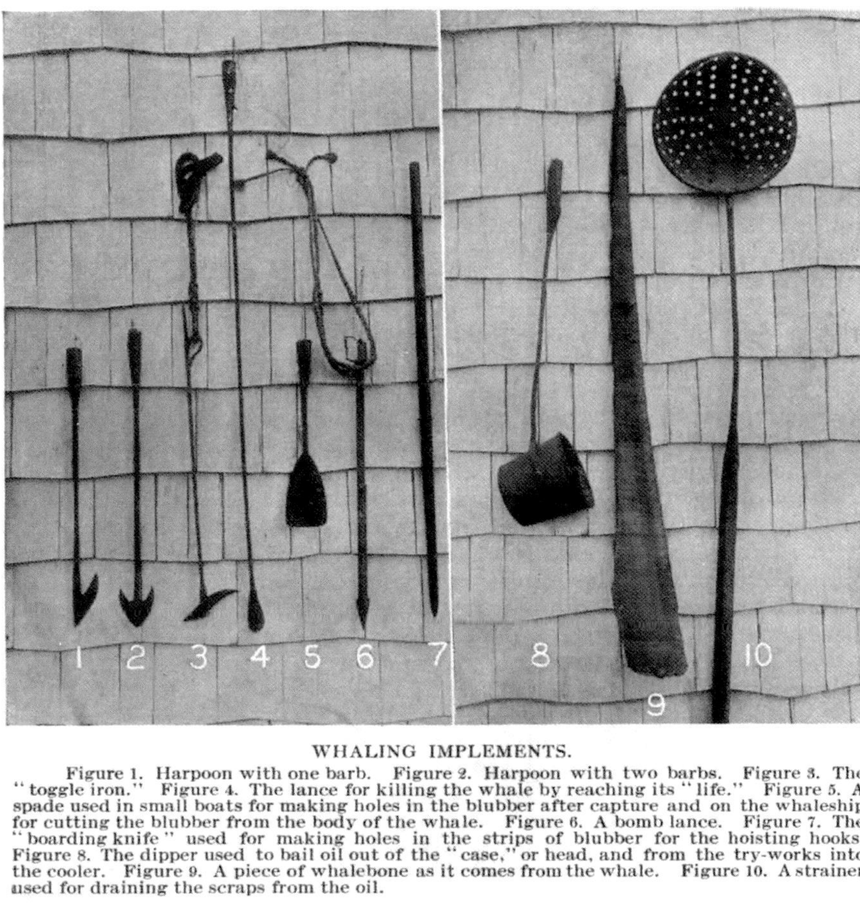

WHALING IMPLEMENTS.

Figure 1. Harpoon with one barb. Figure 2. Harpoon with two barbs. Figure 3. The "toggle iron." Figure 4. The lance for killing the whale by reaching its "life." Figure 5. A spade used in small boats for making holes in the blubber after capture and on the whaleship for cutting the blubber from the body of the whale. Figure 6. A bomb lance. Figure 7. The "boarding knife" used for making holes in the strips of blubber for the hoisting hooks. Figure 8. The dipper used to bail oil out of the "case," or head, and from the try-works into the cooler. Figure 9. A piece of whalebone as it comes from the whale. Figure 10. A strainer used for draining the scraps from the oil.

(Figure 1), shown on the preceding page; another had two barbs (Figure 2); and the third was the "toggle iron" (Figure 3), which has already been described. The edges were sharpened like a razor and were protected by a wooden cover when not in use. They were so sharp that Melville in "Moby Dick" describes his whaling hero, Queequeg, as shaving with one. The lance (Figure 4) which was used after the harpoon had been driven in "to the hitches," or its entire length, resembled a flat spoon, and was very sharp on the edges and on the point. The long line was attached to the harpoon, and shorter lines, called "monkey ropes," were made fast to the lances.

It has been shown by the records of one James Durbee, a veteran harpoon maker of New Bedford, that between the years 1828 and 1868 he made and sold 58,517 harpoons, and he was only one of eight or ten manufacturers of whaling implements in that one port.

An interesting and authentic anecdote of a lost harpoon describes how a Captain Paddock in 1802 struck a whale, which escaped with his iron, and in 1815, thirteen years later, the same captain killed the same whale and recovered his lost weapon.

A whaler is supplied with from four to seven whaleboats, three of which are usually on the port side, one on the starboard side near the stern, and the rest are on deck; it was the improved early canoe, sharp at both ends so as to make a dash at the whale and then be able to retreat just as easily. The floor was very flat so as to enable the boat to be turned quickly in order to dodge a sudden movement of the whale. The boat was about twenty-eight feet long, was equipped with one long steering oar and five rowing oars, and a sail which was occasionally used; also paddles were sometimes resorted to in order to avoid noise. In the bow of the boat two seven-foot harpoons were placed ready for use. A warp was securely fastened to them, and to this warp was secured, after the boat was lowered, a line of two or three hundred fathoms of the best manila two-thirds of an inch in diameter, and with a tensile strength of about three tons. It ran from the harpoons through a chock or groove in the bow to a coil in a depressed box near by, and then lengthwise along the boat to the stout loggerhead or post in the stern, around which it made a turn or two, and then went forward to the line tub near the tub oarsman. Its twelve or eighteen hundred feet of line were coiled in this tub, with every possible precaution to prevent fouling in the outrun. When the rope was coiled and the tub was covered, it was said to resemble a Christmas cake ready to present to the whales. The loggerhead was for snubbing and managing the line as it ran out. A spare line was carried in another tub. A boat was also supplied with extra harpoons, lances, spades, hatchet with which to cut the line if necessary, lanterns, box of food, keg of water, and compass, weighing, all complete, about twelve hundred pounds.

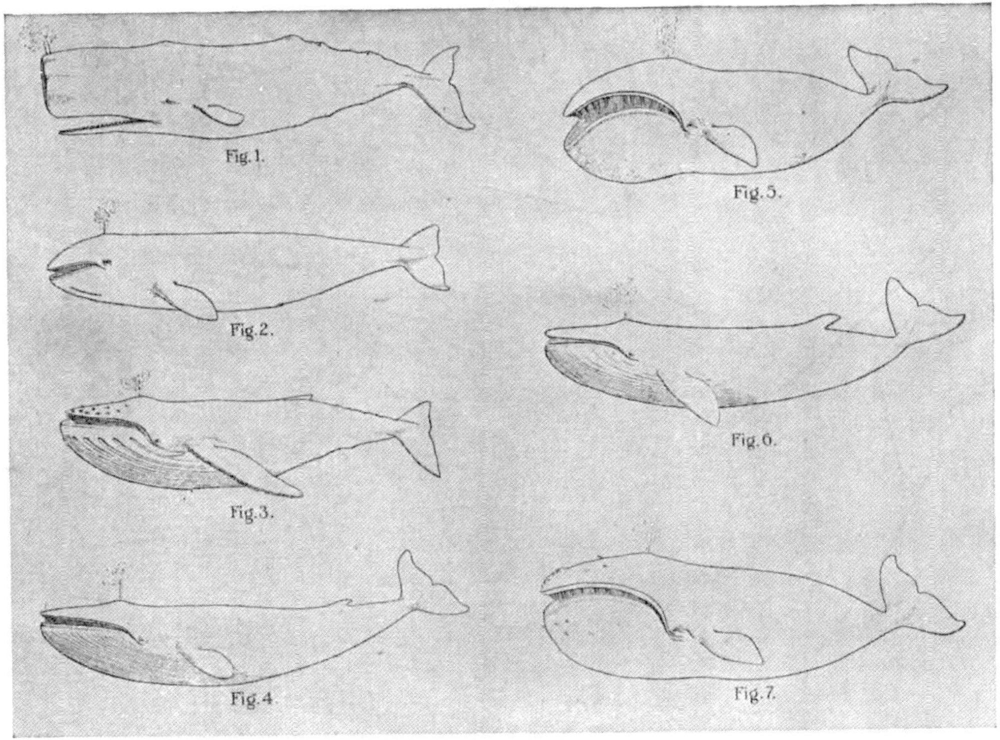

Fig. 1. The Sperm Whale (Physeter macrocephalus).
Fig. 2. The California Gray Whale (Rjachianectes glaucus).
Fig. 3. The North Pacific Humpback (Megaptera versabilis).
Fig. 4. The Sulphur Bottom (Sibbaldius sulfurens).
Fig. 5. The Bowhead (Balaena mysticetus).
Fig. 6. The Finback or Oregon Finner (Balaenoptera velifera).
Fig. 7. The Pacific Right Whale (Balaena japonica).

DIFFERENT SPECIES OF WHALES AND THEIR PRODUCTS

There are many different kinds of whales; namely, sperm whale, right whale, finback, humpback, razor-back, sulphur bottom whale, and the narwhal. The two former species are the more often sought after. The sperm whale was so called because it was the only kind that furnished sperm oil, which is a richer and more valuable fluid than the ordinary whale oil. This species was also called "cachalot." It has one spout hole through which it blows vapor (not water as is generally supposed), which resembles one's breath on a frosty morning; it has also about fifty teeth on the lower jaw which fit into sockets in the upper jaw, and very small eyes and ears. This kind of whale usually employed its mouth as a means of defence, whereas the right whale used its immense tail. A large-sized whale will yield about eighty barrels of oil, but they have been known to boil even larger amounts. Captain John Howland of New Bedford captured two whales which produced over four hundred barrels together. The tongue alone often produced twenty-five barrels. In order to attract the squid, or cuttle-fish, which is often lured by a shiny object from the dark recesses in the great depths of the ocean, the jaw and inner side of the Brobdingnagian mouth are lined with a silvery membrane of phosphorescent whiteness, which is probably the only thing the squid sees when the dark body of the whale is at the great depths to which it sometimes descends for food. Huge pieces of shark and hundreds of mackerel have been found in the stomach of a sperm whale, showing what a carnivorous animal the sperm whale is.

The right whale was so called because it was supposed to be the "right" whale to capture. It differs from the sperm whale chiefly from the fact that it has long strips of whalebone in its mouth which catch the small fish for food, the whalebone serving in place of the teeth of the other species. A right whale usually has about five or six hundred of these parallel strips, which weigh in all about one ton; they are over ten feet long, are fixed to its upper jaw, and hang down on each side of the tongue. These strips are fringed with hair, which hangs from the sides of the mouth and through which the whale strains the "brit," on which a right whale feeds. The "brit" is a little reddish shrimp-shaped jellyfish which occurs in such quantities in various parts of the ocean that often the sea is red with them. With its mouth stretched open, resembling more than anything else a Venetian blind, a sulphur bottom or right whale scoops, at a speed of from four to six miles an hour, through the "brit" just under the surface and thus sifts in its search for food a tract fifteen feet wide and often over a quarter of a mile long. As the whale drives through the water much like a huge black scow, the sea foams through the slatted bone, packing the jellyfish upon the hair sieve. When it thinks it has a mouthful it raises the lower jaw and, keeping the lips apart, forces the great spongy tongue into the whalebone sieve. It then closes its lips, swallows the

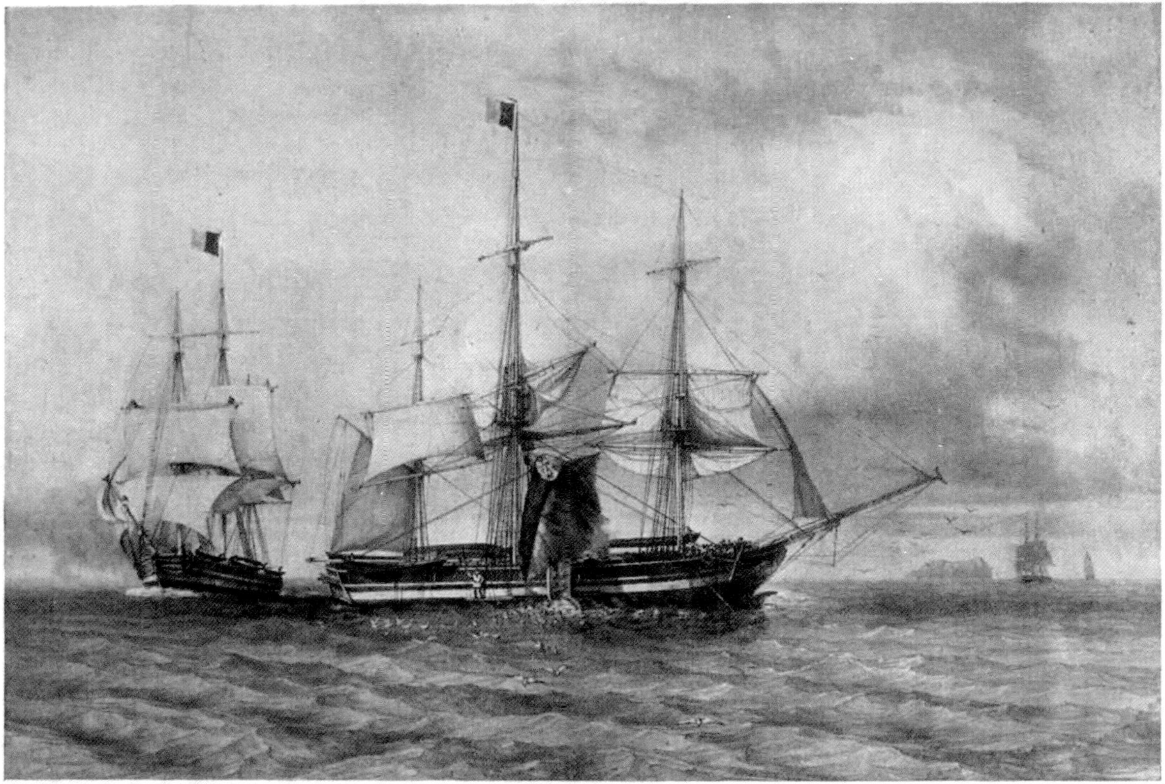

A ship on the northwest coast "cutting in" her last right whale, showing the jaw with the whalebone being hauled on board.

catch and repeats until satiated. Another difference between the sperm and the right whale is that the latter has two spout holes instead of one.

The sperm whale is found in the warm waters off the coasts of Chili, Peru, Japan, New Zealand, Madagascar, California, and Brazil; in the Caribbean, China, and Red Seas, in the Indian Ocean and Persian Gulf; off the Azores, Java, Galapagos, Society, Sandwich, Fiji, and Samoan Islands; and off the Cape de Verdes. The right whale is found in the high latitudes of the Arctic Ocean, in Baffin's Bay, in the Ochotsk Sea, near Tristan d'Acunha and the Desolation Islands, and in the Japan Sea. There were many other cruising grounds, but these were the most frequented.

The finback is even longer than the other varieties, but whalers rarely attack it owing to the thickness of the blubber and also owing to the fact that it swims so fast that, to use a favorite expression of whalemen, it "will run the nails out of the bottom of the boat."

The "narwhal," or nostril whale, has a horn five to ten feet long protruding forward from its jaw. This species is also spoken of as the "Unicorn." Opinions differ as to the use of this horn; some think it is used as a rake to turn over its food at the bottom of the sea, others think it is employed as an ice-piercer, but the author of "Moby Dick" suggests that it would make an exceedingly good folder for it to use in reading pamphlets. In ancient times this narwhal's tusk was used to detect poison in food and wine, the idea being prevalent that the tusk would be discolored if it came in contact with any poisonous substance. It is difficult in the present day to appreciate the wholesale fear of poison which existed up to quite modern times. This fear was so general and pressing that no one of any position dared to eat and drink without a previous assurance that what was set before him did not contain some poison. Some authorities vouch for the fact that the tusk was also used as salts for fainting women.

The chief products of the fishery are sperm and whale oil, whalebone, and ambergris. Spermaceti, meaning a foot of "sperm oil," was the most valuable and was found only in the sperm whale. This oil was formerly used chiefly in the manufacture of sperm candles, and at one time there were eight factories for the manufacture of these candles in New England, Nantucket alone turning out three hundred and eighty tons annually before the war. In the olden times this oil was considered a sure cure for almost any kind of disease and was worth its weight in silver. Shakespeare makes reference to it in these words—"The sovereign'st thing on earth was 'parmaceti for an inward bruise." At present it is used chiefly in making refined oils for lubricating.

Whale oil was procured from all the other varieties of whales, and was formerly used as an illuminant in the old "whale oil" lamps; it is used now to a certain extent in the tanning of leather and in the manufacture of soaps, but chiefly in making heavy lubricating oils.

Whalebone has been the most important product of the whale fishery for a number of years, and in fact whaling would undoubtedly have died out altogether had it not been for the discovery of its use in making women's stays. Many a whaleman has lost his life in the endeavor to improve the female figure. It is a curious fact that fifty years or more ago this product was always thrown away as worthless. The value has gone down in the past few years on account of the invention of steel stays, which take the place of whalebone.

The high and low prices of these three commodities are of interest. Sperm oil was $2.55 per gallon in 1866, and is 46 cents now. Whale oil was $1.45 per gallon in 1865, and is 26 cents now. Whalebone was $5.80 per pound in 1904, 8 cents in 1809, and is $1.75 now.

Ambergris, the rarest and most valuable of all the products, is a secretion from the intestines of the sperm whale and results from a disease. It is a very rare article and is worth almost its weight in gold, selling usually at $300 a pound. Its chief use is in the preparation of fine perfumeries. It is believed that the largest amount taken by one ship was brought back by the "Watchman" of Nantucket, which vessel found eight hundred pounds in 1858. Small amounts were sold every year in New Bedford even up to the year 1913. The Turks used it in cooking and also carried it to Mecca for the same purpose that frankincense is carried to St. Peter's in Rome. Some wine merchants used to drop a little into their wine as a spice, and it was said that the Moors used it in green tea as a flavoring to present to their guests.

The whale is used for food chiefly by the Japanese and Esquimaux, and a famous doctor belonging to the latter tribe some years ago recommended the blubber for infants. In fact, the whale would perhaps be considered a good dish were there not so much of him. Whalemeat is said by some to resemble boarding-house steak. In France, during the Middle Ages, the tongue was considered a great delicacy, and by some epicures the brains, mixed with flour, were much sought after.

The largest income received by the whalers of America in any one year was in 1854, when they netted $10,802,594.20, although the record size of the fleet was attained eight years before. The five years from 1853 to 1857 inclusive yielded a return of $51,063,659.59, the catch of each year selling for fifty per cent. of the total value of the whaling fleet. The total value of the cargoes from 1804 to 1876 was $331,947,480.51.

Captain W. T. Walker, of New Bedford, is called the counting-house hero of the American Whale Fishery. He purchased in 1848 an old whaleship called the "Envoy" that was about to be broken up, and when ready for sea this ship stood the owner $8,000. He could get no insurance; nevertheless he "took a chance," and after a three years' voyage he returned and had netted for himself the extraordinary sum of $138,450, or 1,630 per cent. The largest profit, however, was made by the "Pioneer" of New London, in 1865, the value of her cargo being $151,060. For a short voyage Frederick Fish, who has been mentioned before, holds the record for his ship the "Montreal," which brought back a cargo worth over $36,000 after a voyage occupying only two months and fifteen days.

There were many unprofitable voyages, and many were the ships that came home with barrels filled with salt water instead of oil for ballast. Some vessels, as whalemen say, didn't have enough oil to grease their irons.

METHODS OF CAPTURE AND "TRYING OUT"

"Whales has feelin's as well as anybody. They don't like to be stuck in the gizzards an' hauled alongside, an' cut in, an' tried out in those here boilers no more'n I do!" *Barzy Macks's Biology.*

When the lookout at the masthead shouts out "Thar she blows," or "There she whitewaters," the whaleboats are gotten out and rowed towards the whale, while signals from the ship show from time to time the whereabouts of the whales and directions for their pursuit. The first man to "raise oil"—an expression which means the first to see a whale—usually received a plug of tobacco or some other prize, and this made the lookouts more keen.

In "Moby Dick" Melville says that the crew pulls to the refrain "A Dead Whale or a Stove Boat," which became such well-known by-words among whalemen that when Mr. W. W. Crapo last year presented to New Bedford "The Whaleman" statue, they were inscribed upon it. When rowing in a rough sea the captain cautioned the men to trim the boat and not to "shift their tobacco."

As they approach the whale the bow oarsman, who is the harpooner, stands up at a signal from the captain of the boat, who is steering, and yells out to "give it to him." The next order is probably to "stern all" in order to avoid the whale. The boat is probably now fast, and either the whale will sound and run out the line at a terrific rate or else he may race away dragging the boat after him, which whalemen call "A Nantucket Sleigh-Ride." This kind of sleigh-ride was often at railroad speed and was perhaps one of the most exhilarating and exciting experiences in the line of sport. An empty boat would certainly capsize, but a whaleboat had six trained, strong, athletic men sitting on her thwarts, whose skill enabled them to sway their bodies to the motions of the boat so that she would keep an even keel, even though her speed might plough small valleys over the huge swells and across the broad troughs of an angry Pacific, and great billows of foam piled up at her bow while the water rushed past the stern like a mad whirlpool. The greatest care must be taken not to allow the line to get snarled up or to let a turn catch an arm or leg, for it would result in almost immediate death to the person thus entangled. Conan Doyle, who once took a trip on a whaler, tells of a man who was caught by the line and hauled overboard so suddenly that he was hardly seen to disappear. One of the men in the boat grabbed a knife to cut the line, whereupon another seaman shouted out, "Hold your hand, the whale'll be a good present for the widow!"

There is one case known where a man who had been hauled down by the line had the presence of mind to get out his knife and cut the rope,

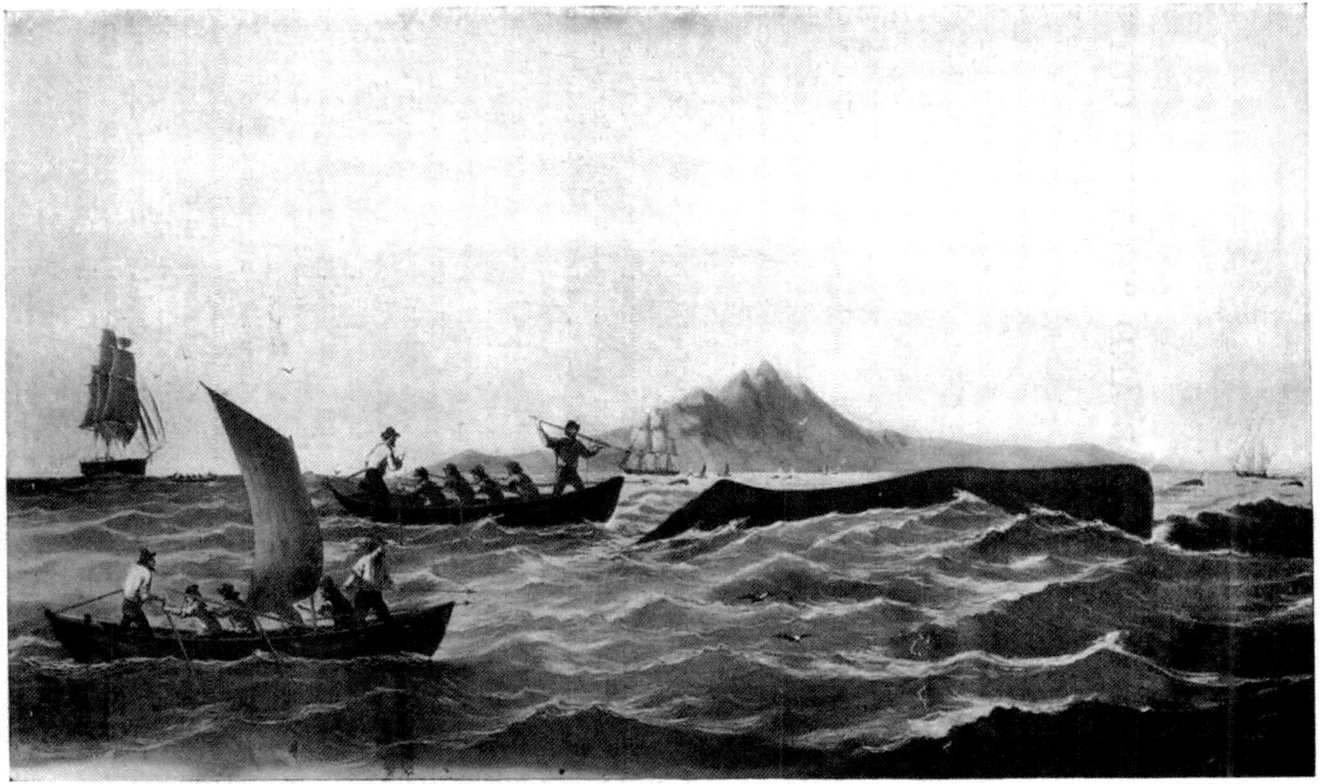

No. 1. "The Chase." A rare New Bedford print.

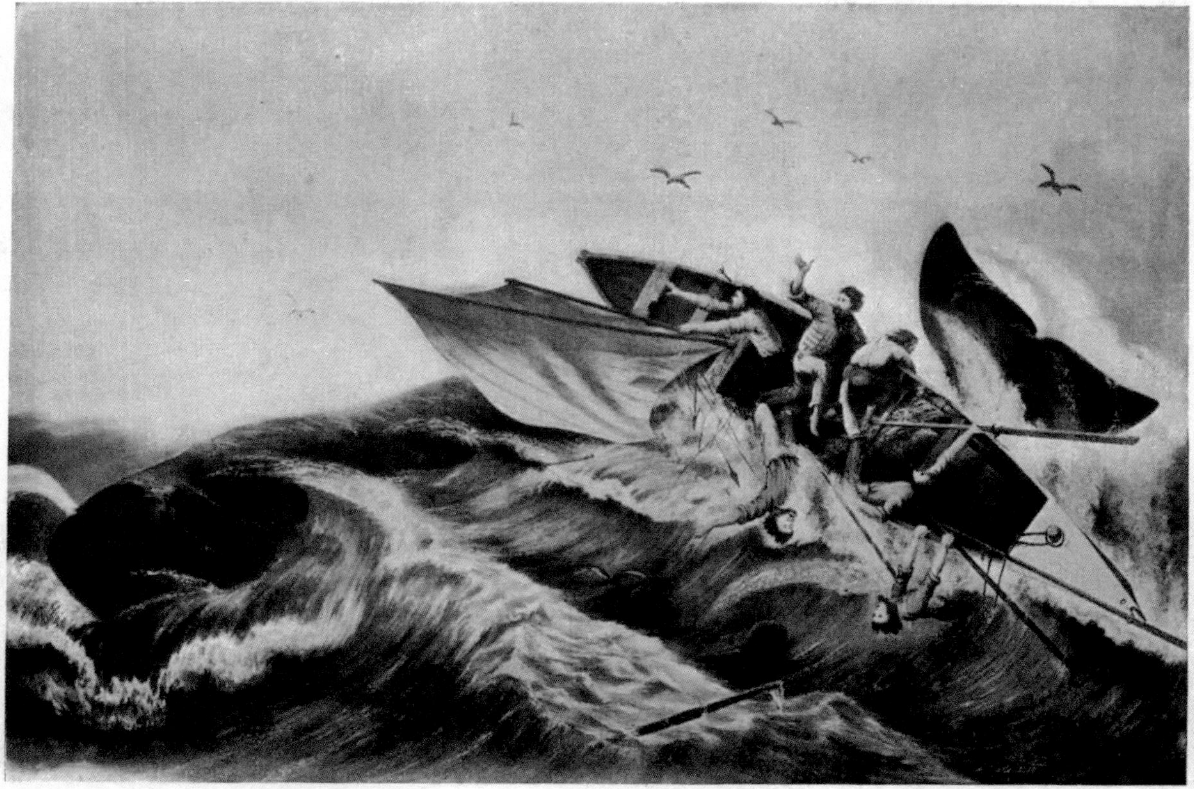

No. 2. "The Conflict," showing ratchet in bow through which the line is run, and post in stern around which line is placed.

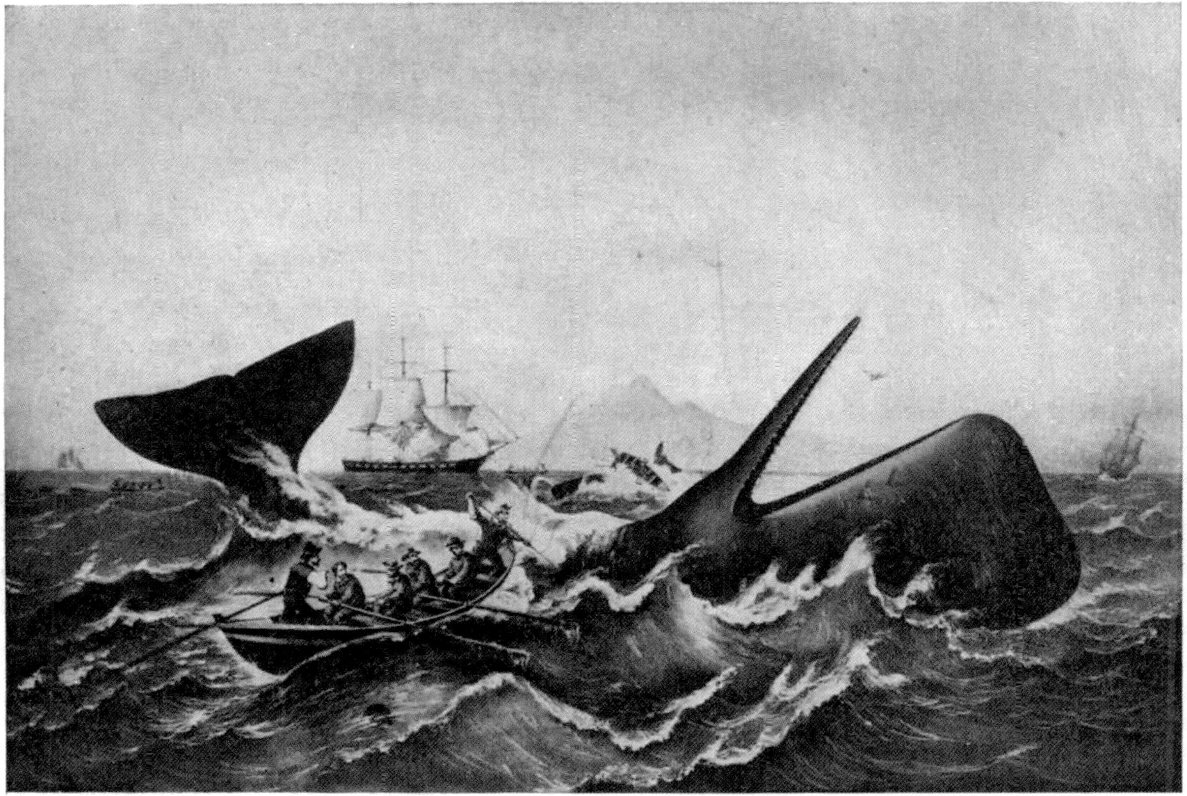

No. 3. "The Capture." A whale will usually turn on its back when dying.

which allowed him to come to the surface more dead than alive; also occasionally the entangled arm or ankle would be torn off, thus freeing the man and allowing him to rise.

Two harpoons were thrown if possible, and then it was customary for the harpooner to exchange places with the boat-steerer, who got ready his lance, which he plunged in and hauled out again until the whale went into his "flurry" and rolled over dead, or "fin out" as it was called. Often the whale would get frightened or "gallied," or would jump in the air or "breach," and therefore great care was taken to avoid his attacks. When the whale "breaches" the tail becomes very conspicuous, and one old salt used to say that an additional tail appeared after every glass of grog.

Scoresby speaks of a whale which drew out from the different boats ten thousand four hundred and forty yards, or nearly six miles, of rope. It was necessary when the line of one boat was nearly exhausted to bend on the end to a new rope in another boat and so on, and of course often miles of rope and many harpoons would be lost if the whale escaped. When the line was drawn out rapidly it was necessary to pour water over the snub post to keep the rope from burning.

There have been races almost as exciting as a Harvard-Yale race when the boats of different nations have been dashing for a whale, which is prized at between three thousand and four thousand dollars. Many years ago an English, a French, a Dutch, and an American ship lay becalmed in the Pacific, when suddenly a whale was "raised." All four ships lowered and raced across the waters, with the American in the rear. In a few minutes the Yankee passed the Dutchman, who yelled "donner und blitzen!" The American captain encouraged his men by shouting "Thar she blows, she's an eighty-barreler, break the oars, lads!" and soon the French were left astern with curses of "Le diable." The Englishmen were still ahead; the American boat-steerer now began to help the stroke oarsman by pushing his oar, and their boat crept up slowly upon their only rivals. The English boat-steerer also grabbed his stroke's oar, but it snapped off at the rowlock, and the Americans overtook them and captured the whale. Another international race took place in Delagoa Bay, which has become a classic among American whalemen. Again an English and a Yankee whaleboat were chasing a whale, and, in some manner, the former was able to cut in between the whale and the Americans, and as the English harpooner was reaching for his iron, the American harpooner "pitch-poled" his harpoon over the English boat, and his iron made fast.

After a capture came the long, hard row back to the ship, then the tedious process of "cutting in" and "trying out." First of all the head, or "case," was cut off and tied astern while the strips of blubber were cut from the body and hauled on board, as next shown, by means of huge tackles from the mast. Blubber averages in thickness from twelve to eighteen inches, and if cut four and one-half inches thick would carpet a room sixty-six feet long by twenty-seven wide. Then the head was either bailed out, if it were a sperm whale, or else

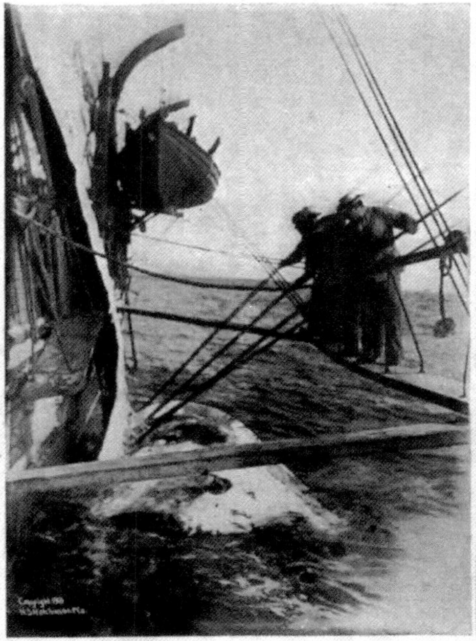

A "cutting" stage, showing blubber being stripped from the whale.

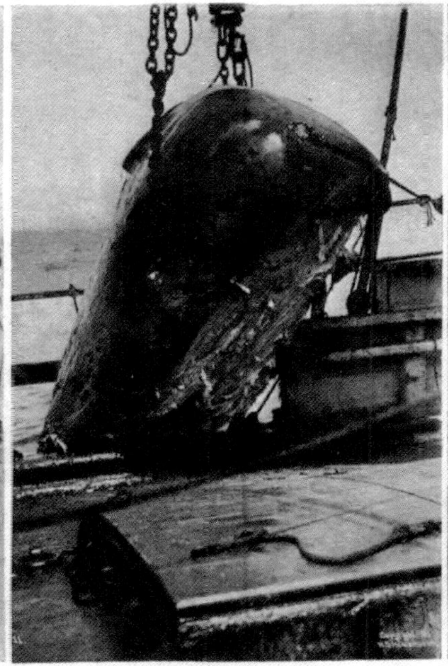

Hauling the "case," or head, on board. The case weighs sometimes as much as 30 tons.

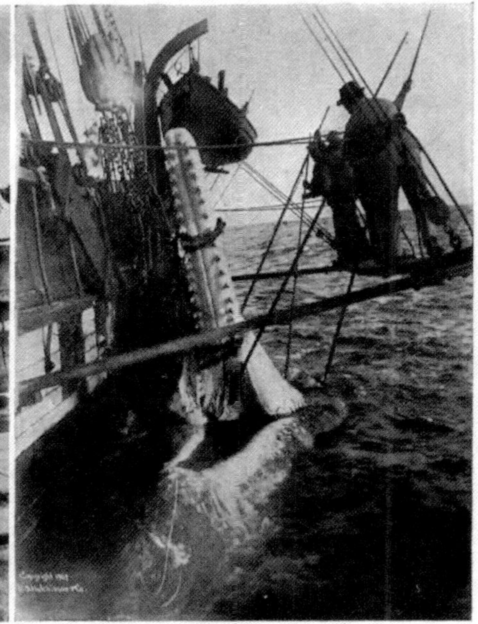

Cutting off the lower jaw of a sperm whale, showing the teeth.

the whalebone was taken in, if it were a right whale. The strips or "blanket pieces" were then minced, and after boiling, the oil was cooled and stored away in barrels below deck. The "try-works" consisted of iron pots set in brick furnaces, and there were pans of water underneath to prevent the decks from burning. This process of boiling the oil was most irksome and disagreeable as the men were soaked in oil from head to foot, and the smell of the burning fluid was so frightful that it has often been alluded to as Hell on a large scale, and was usually called a "squantum," which is the Nantucket word for a picnic; nevertheless, old whalers delighted in it.

It is a superstition among some whalemen that a ship which for once has a sperm whale's head on her starboard quarter, and a right whale's on her port side, will never afterwards capsize.

THE PERILS OF WHALING

Whalemen not only had to undergo the perils of the sea, but in addition ran the danger of being killed by the whale and of being attacked by savages at the ports where it was often necessary to land for food and water. Also in cases of accident the whaleship was usually off the regular cruise followed by the merchantmen and therefore less likely to be assisted by other vessels. Furthermore, the long voyages, poor food, and the many dangers of whaling induced many mutinies.

The worst massacre occurred on the "Awashonks," of Falmouth, in 1835, near the Marshall Islands. The natives came on board in large numbers and seemed most friendly, when, on a given signal, they killed the captain and many of the crew. Finally the seamen laid a charge of dynamite under a hatchway where the savages were sitting, and blew most of them to pieces, the crew being then enabled to recapture the vessel. A few years later, when the "Sharon" of Fairhaven was cruising not far from Ascension Island, the crew lowered for a whale, and upon returning to the ship it was discovered that three of the "Kanaka" crew, recently engaged, had taken charge of the ship and had killed the captain. The first mate in the whaleboat did not dare attack, but the third mate, Benjamin Clough, who was only nineteen years old, swam to the ship in the darkness, climbed up the rudder, shot two of the mutineers, and had a hand-to-hand encounter with the third, who died soon afterwards. The first mate then returned on board. Clough was made captain of a ship immediately upon his return to Fairhaven. Still another mutiny took place on the ship "Junior" which sailed from New Bedford in 1857, most of the officers being killed. Plummer, the ringleader, wrote a story of the mutiny in the log book, which is now in the possession of the New Bedford Library, and the account was signed by the five mutineers in order to clear the rest of the men on board. The five murderers on sighting land lowered two whaleboats with all the plunder they could find and rowed ashore. The mutineers were subsequently captured and were brought in cages to Boston, where they were defended by

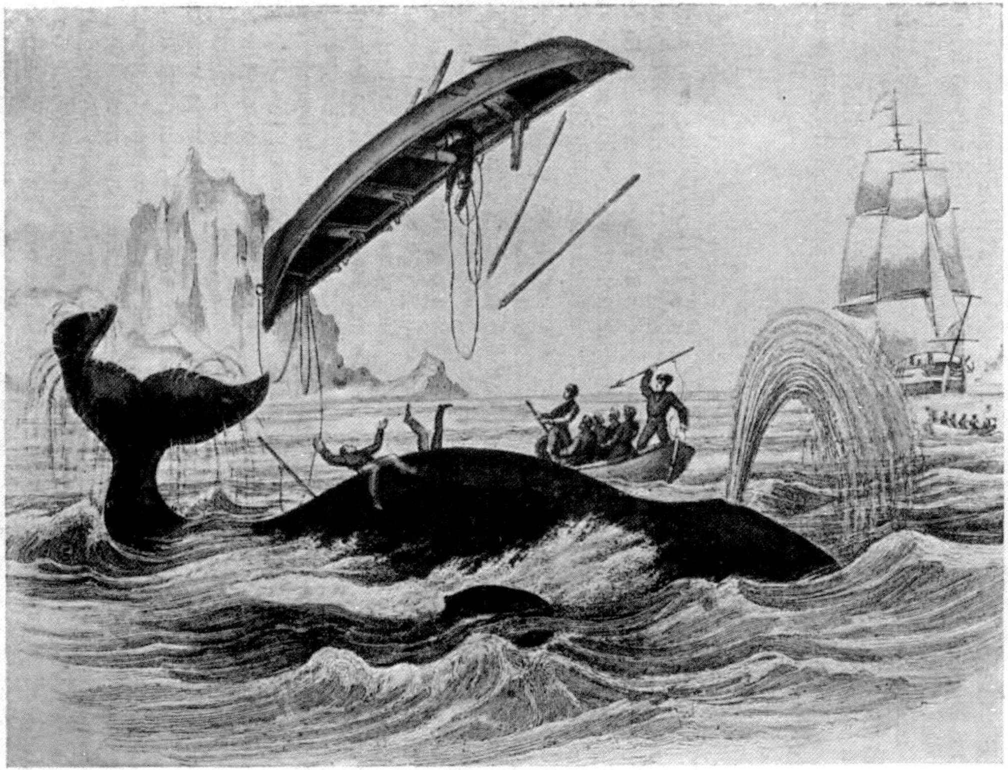

A whale playing battledore and shuttlecock with a 1200-pound whaleboat and six men.

Benjamin F. Butler. Davis, the author of "Nimrod of the Sea," mentions a quarrel on board the "Chelsea," which ended by the men all signing a "round robin" to return to duty, and in order that no name should head the list the signatures were set down in a circle, like the spokes of a wheel, from which possibly comes the word "ringleader."

The most fearful mutiny happened on the "Globe" of Nantucket, in 1822. A boat-steerer called Comstock laid a plot which resulted in the death of all the officers of the ship, and those who were not killed outright were thrown overboard. Comstock then took charge of the ship, and stated that if any man disobeyed him, he would be put to death by being boiled in the "try-pots." The ringleader was finally killed by some of the crew, and the ship brought into port.

Captain Warrens, of the whaler "Greenland," in 1775, told a most thrilling narrative, which shows the perils of Arctic whaling, and is the most weird and grewsome of all whaling yarns. While becalmed one day he sighted a vessel with rigging dismantled, and he immediately lowered and rowed over to her. Upon boarding the ship he found seated at the cabin table the corpse of a man. He held a pen in his hand, and the log book was on the table in front of him. The last entry was "Nov. 14, 1762. We have now been enclosed in the ice seventeen days. The fire went out yesterday and our master has been trying ever since to kindle it again without success. His wife died this morning. There is no relief." Other corpses were found in the cabin and a number of sailors in the forward part of the ship. The vessel had been frozen in the ice for thirteen years!

There are many exciting accounts of accidents to whaleboats, and a few are worth mentioning. Captain Sparks, of the "Edward Lee" of Provincetown, in 1881, chased a whale and finally lost him. He and his crew endeavored to find his ship, but for some reason were unable to do so. The nearest land was one thousand miles away, and with no food or water the prospect was not very encouraging. For six days they sailed on, when by good fortune they killed a whale, and finally were picked up and brought to land.

Another incident shows how a whale will sometimes fight. Captain Morse, of the "Hector" of New Bedford, had his boat attacked by a whale, which grabbed the bow in its mouth, shaking the crew and implements in all directions. The mate came to the rescue, and the whale at once started to chase his boat, snapping its jaws less than a foot behind the stern. The crew rowed desperately and succeeded in dodging its attacks, until finally the animal turned over to get more air, and a well-driven lance luckily killed it. The harpoons of the "Barclay" were found in it, and it was learned that this same whale had killed the "Barclay's" captain only three days before. Another incident shows the fierceness of the attack of a fighting whale. The "Osceola 3rd," of New Bedford, shot thirty-one bombs into a whale before it was killed.

Captain Davis, in "Nimrod of the Sea," mentions an occurrence in which a whale attacked one of the men who had been hauled from the

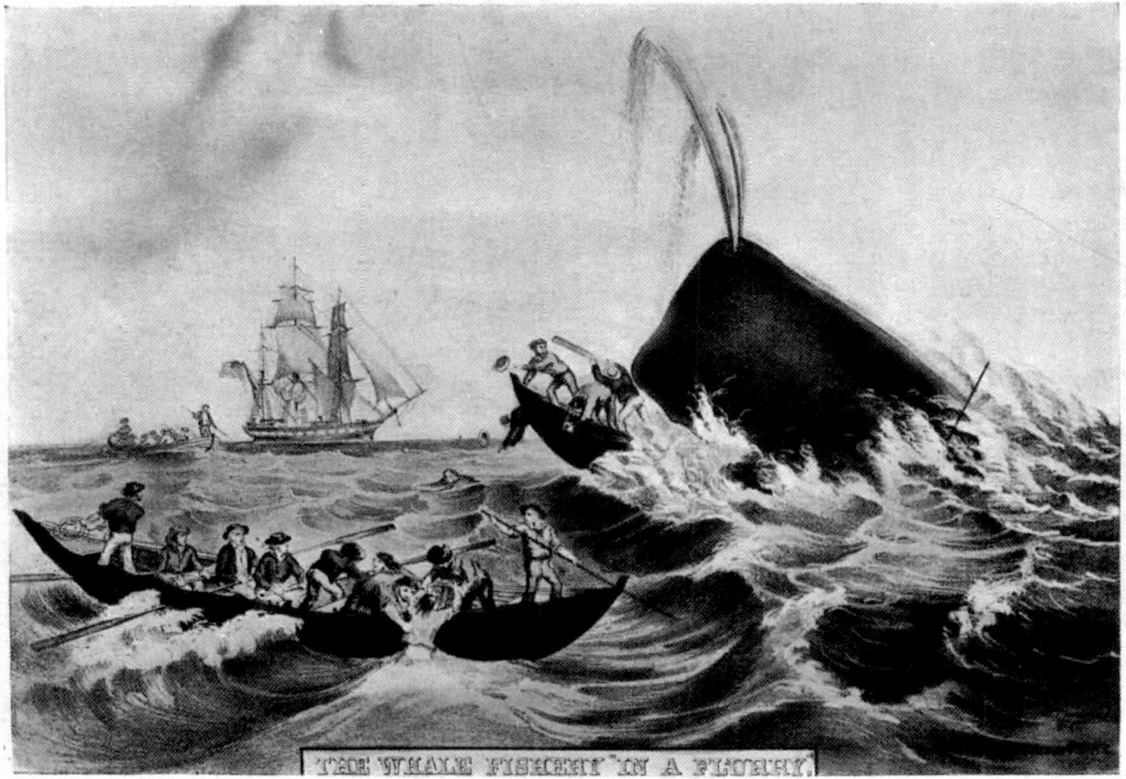

A whale is often fond of eating whaleboats and men.

whaleboat. Then ensued a fight, and every time the monster swam for him he was obliged to dive. The mate rushed into the encounter with his boat and finally succeeded in killing the whale. Another captain described how the crew of his whaleboat was obliged to cling all night on the body of a dead whale until help came at daybreak. It happened to be Christmas evening, and the famished men obtained their Christmas dinner by digging from the back of the dead animal enough meat to satisfy their hunger. If a whaleboat were upset, and it was seen that the crew had something to hold on to in order to prevent going under, it was often a long time before the other boats rendered assistance, it being a truism among whalemen that whales were of much higher commercial value than men.

Captain Hosmer, of the bark "Janet" of Westport (near New Bedford), met with a horrible experience off the coast of Peru in 1849. He had just secured a whale, and in towing it back to his ship his boat was capsized. He immediately displayed distress signals, and the "Janet" sailed towards the men who clung to the small boat, when suddenly, to his amazement and horror, the ship swung off and headed in another direction. They could see her sailing about searching for them, but were unable to attract her attention, and finally, as the distance between them increased, they set sail towards the nearest land, after bailing out their boat with difficulty, and having lost one man by drowning. The nearest coast was over one thousand miles away, and they had not a drop of water or a morsel of food. At the end of seven days lots were cast to decide who should be killed in order that the rest might live. Four more of the crew died, and after twenty days the two survivors landed on an island and were later picked up by the "Leonidas" of New Bedford.

There are three cases known to history of a whale sinking a ship. The "Essex," of Nantucket, was attacked by a huge whale in 1819, and twice did the animal make a rush at the ship, which became submerged in a few minutes. Owen Chase, the first mate, wrote an account of the accident and subsequent sufferings of the crew. Three whaleboats set sail for the Marquesas Islands. One boat was never heard from; another was picked up by an English brig with only three of the crew alive; and the third with only two survivors, having sailed over twenty-five hundred miles, was picked up by a Nantucket vessel, *three months* after the accident. Captain Pollard, who was in command of the "Essex" at this time, had previously been one of the crew on Fulton's "Claremont" on his first trip up the Hudson. He survived the frightful experience, but nothing could induce him ever to refer to it. He finally abandoned the sea and became a police officer in Nantucket.

The "Ann Alexander" of New Bedford, which is shown in the next cut, met a similar fate in 1850, and the ship sank so quickly that only one day's supplies were saved. With the horror of the "Essex" staring them in the face the crew set sail in the small boats, and with great good fortune in two days sighted the "Nantucket" and were taken

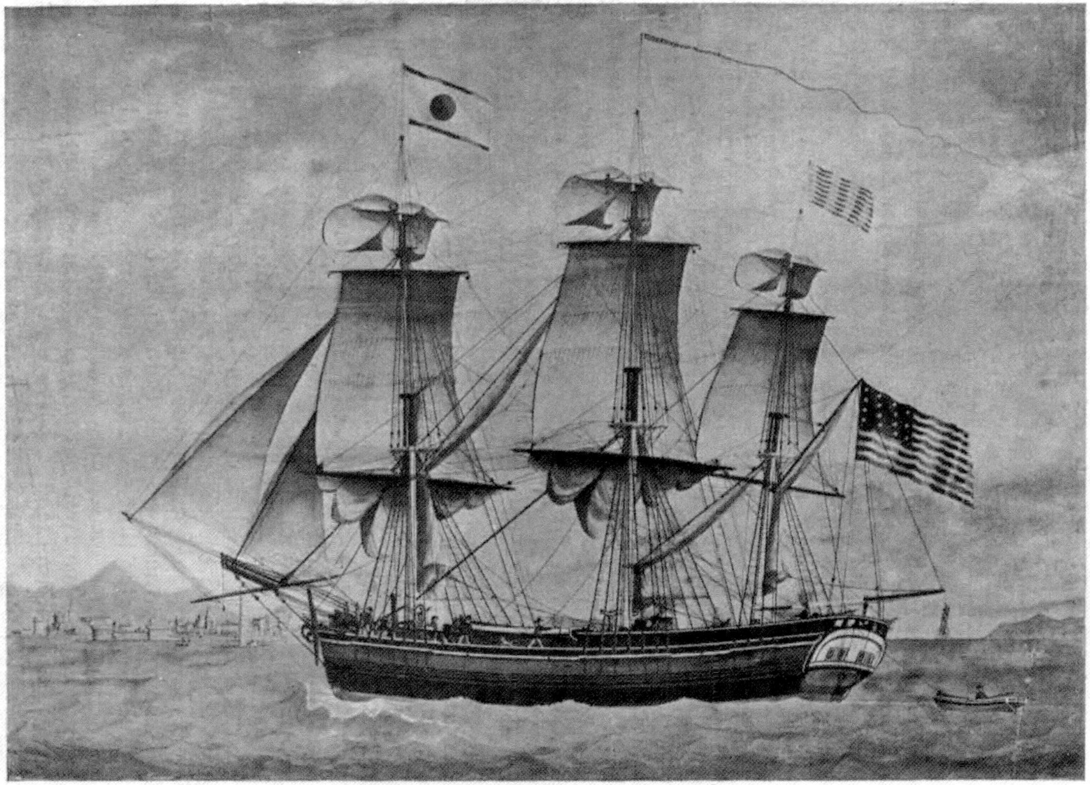

The "Ann Alexander" of New Bedford.

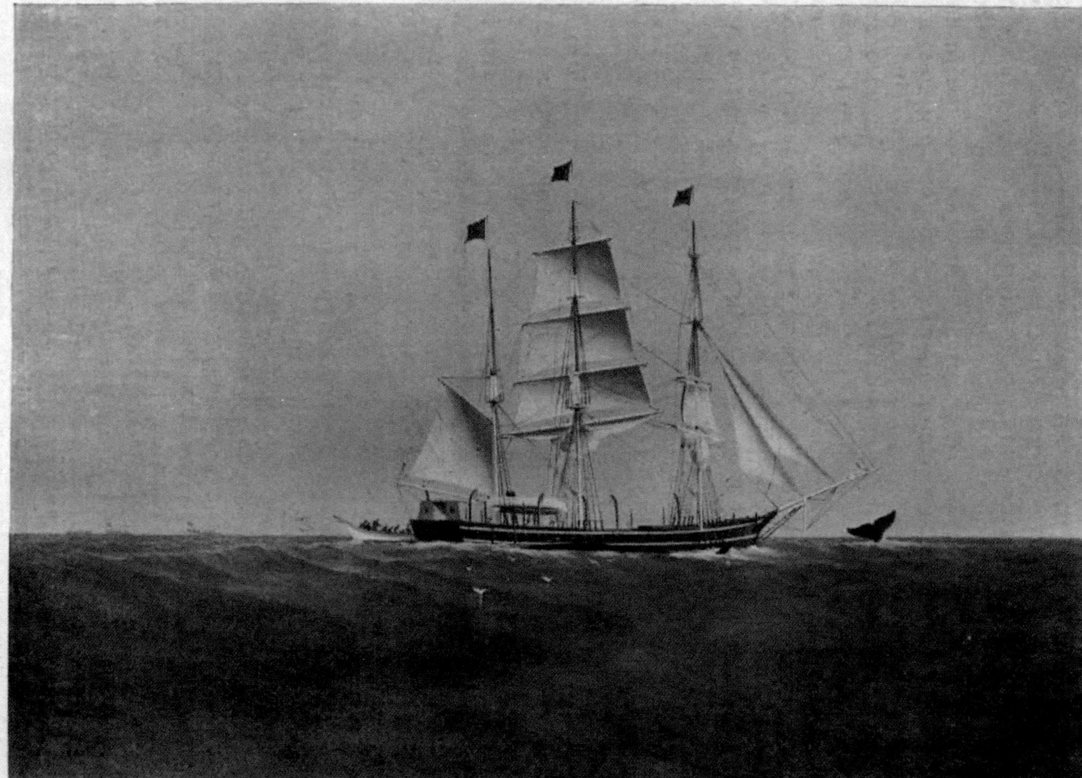

The "Kathleen" of New Bedford sinking in mid-ocean, having been "stove" by a monster whale. Flags at the mastheads are signals for the three whaleboats to return.

on board. Five months after this incident the "Rebecca Sims," of New Bedford, killed a whale, and to the great surprise of the crew, the irons of the "Ann Alexander" were discovered in its body, and there were also several pieces of the ship's timber imbedded in its head.

The latest of the three accidents happened to the bark "Kathleen" in the Atlantic Ocean in 1902, and the picture shows her about to sink after having been rammed by a whale. The three flags at the mastheads are signals to the three boats to return at once, but as each one was fast to a whale, they were loath to obey the signals. The whale showing its "flukes" at the right of the picture is the one that stove the hole in the vessel. The "Kathleen" also had a whale alongside, making four just captured. The accident meant a loss, not counting the vessel and oil on board, of ten to twelve thousand dollars. Captain Jenkins, who was in command, lowered with Mrs. Jenkins, a parrot, and nineteen of the crew, and with difficulty rowed to the other boats, which took in their share of the men from the captain's overcrowded one. Captain Jenkins declares that the parrot, when removed from its home on the "Kathleen," swore that "he would be damned if he'd ever go to sea again!" Three boat loads were discovered by a Glasgow ship, but the fourth had to sail over one thousand miles to the Barbadoes. Captain Jenkins is to-day living in South Dartmouth. He has written a small volume on the loss of his ship and is such a well-known whaleman that he was one of those who occupied the platform at the time of the unveiling of "The Whaleman" statue.

THE "CATALPA" EXPEDITION

While not primarily a whaling voyage, the "Catalpa" Expedition should be outlined in any account of whaling adventures.

A number of Irish subjects who had joined the Fenian conspiracy of 1866 had been banished to Australia for life and were serving in the English penal colony at Freemantle. John Boyle O'Reilly had escaped with the aid of a whaleship and immediately began to form a plot to release his fellow prisoners. O'Reilly suggested a whaleship for the rescue, chiefly because it would create little suspicion, as whaleships were frequently seen off the coast of Australia. Captain H. C. Hathaway, who was the head of the night police force at New Bedford, was then consulted, and he recommended their approaching a certain George S. Anthony, a most successful whaler. Accordingly a meeting was held in a dark room, and Captain Anthony finally accepted the leadership of the expedition, probably not realizing fully the danger involved. The "Catalpa" was selected, and she sailed from New Bedford on April 29, 1875, not even an officer sharing the secret with the brave commander. The ship actually captured whales and finally arrived off Bunbury on the coast of Australia. In the mean time a man called John J. Breslin, who used to be a freight agent in Boston, had gone to Australia with a fellow conspirator to arrange the land end of the scheme. On the day appointed Captain Anthony rowed ashore with

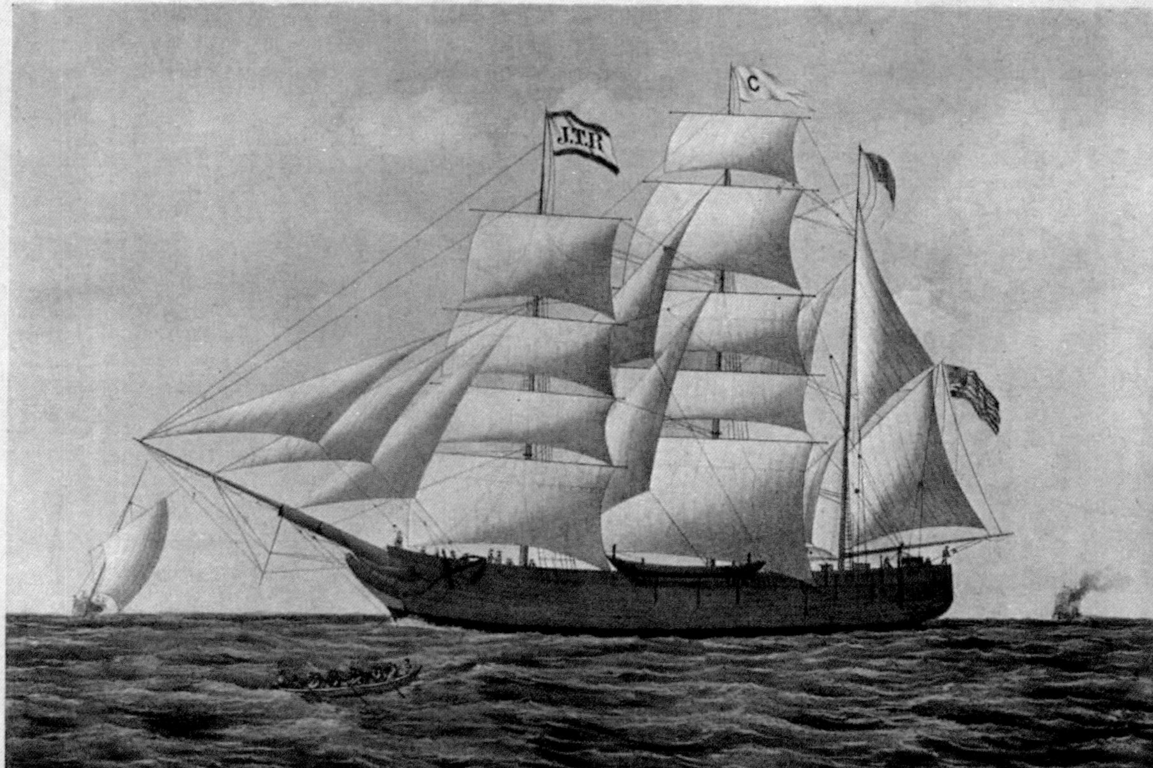

Whaling-bark "Catalpa" of New Bedford rescuing prisoners from Australia in 1876; on the left is the police-boat racing to intercept the convicts in the rowboat, and on the right is the English armed cruiser "Georgette" coming to the assistance of the police. The prisoners barely escaped.

his crew, and with great difficulty Breslin and his six prisoners, who had escaped from their work in the woods, were placed on board the rowboat, which set out to sea to join the "Catalpa," some miles off shore. A storm came up, but by good fortune and skilful seamanship, after a whole day and night, the "Catalpa" was sighted. At the same time the English cruiser "Georgette" was seen coming out of Freemantle in search of the refugees. By great luck for some reason she never noticed the small whaleboat and after questioning the "Catalpa" put back towards the shore. The rescued and rescuers rowed on and finally were observed by the men on the "Catalpa." At the same time Captain Anthony noticed with horror that there was an armed guard boat almost as near the "Catalpa" as was his boat. It was a terrific race, but the whaleboat arrived a few seconds ahead and the occupants climbed on board; the officers had lost, and the prisoners were free. The rescued men knew their pursuers and, leaning over the rail of the "Catalpa," wished them "Good morning," and there was nothing for the officers to do but to answer them in the same tone. When the captain reached home he weighed one hundred and twenty-three pounds, having lost thirty-seven pounds on the voyage, through worry and excitement. The police of Western Australia endeavored to get these prisoners returned, but as their letter was addressed to the same Captain Hathaway who assisted the plotters of the expedition, there was not much help in this direction!

It is a very curious fact that at the precise moment that Disraeli was telling the House of Lords that he would not release these prisoners they were free on the Yankee ship. Receptions were held in New Bedford and Boston in honor of Captain Anthony and the other rescuers, and the daring captain will always be a hero with the Irish people.

DECLINE OF WHALING AND THE CAUSES

The first whaler to sail from San Francisco was the "Popmunnett" in the year 1850, and for thirty years after there were a few whaleships registered in this port. Steam whalers were introduced into the American fleet in 1880, when New Bedford sent out one, but it was the adoption of steam and the proximity to the Arctic that made San Francisco a whaling port at the time other places were giving up the pursuit. In 1893 there were thirty-three vessels enrolled there, many of which had been transferred from the Eastern cities. Since 1895 Boston, New Bedford, Provincetown, and San Francisco have been the only places from which whalers have been regularly registered, and in 1903 Boston recorded her last whaleship.

There are a number of reasons for the decline of the whale fishery, but the chief factor was undoubtedly the introduction of kerosene. The opening of the first oil well in Pennsylvania sealed the fate of whaling. Henceforth sperm candles were used for ornament, and whale oil lamps soon became interesting relics. Other causes doubtless con-

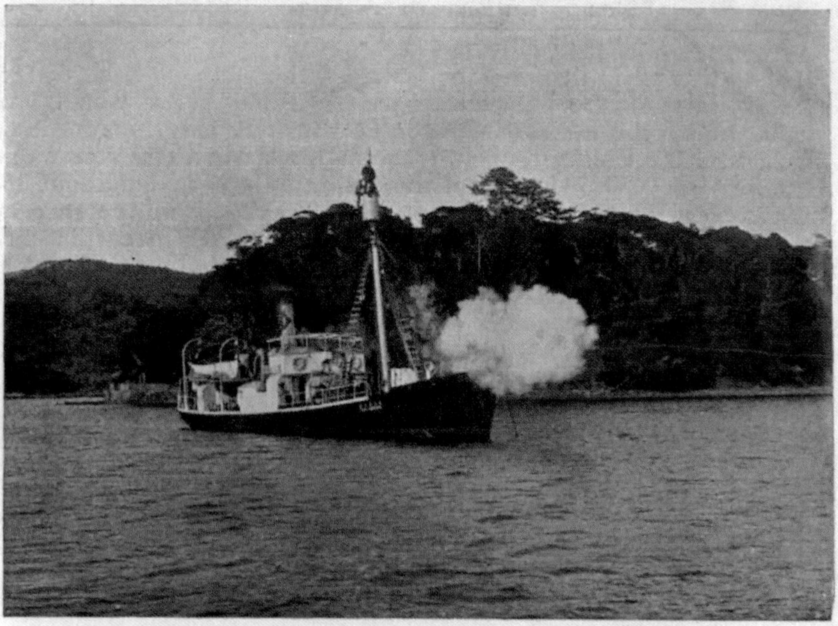

A modern steam whaler in the act of shooting a harpoon gun.

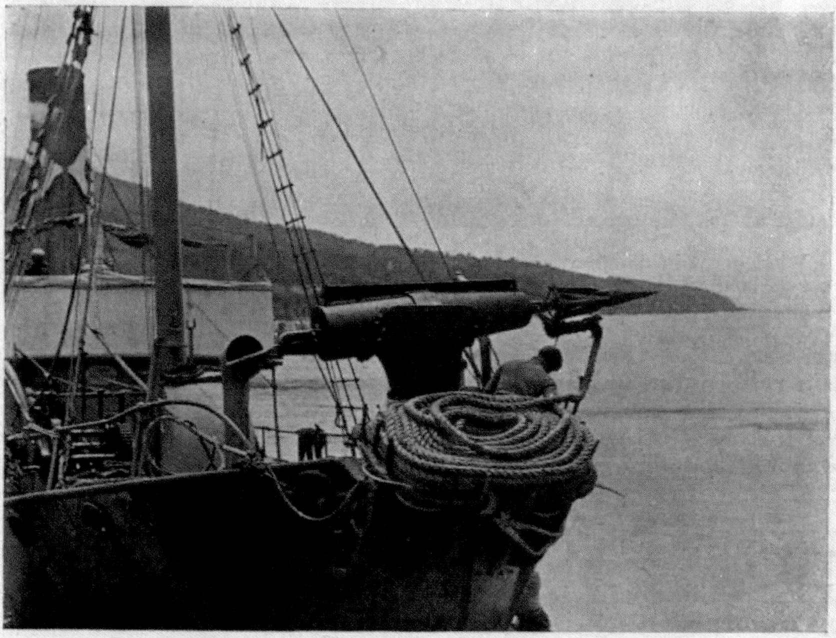

The modern harpoon gun, showing line with which to hold the whale.

tributed to this rapid decline; for instance, the financial crisis of 1857; the uncertainty of the business, especially since Arctic whaling was begun in 1848; the increased cost of fitting out the ships for longer voyages; and the California gold craze in 1849, when many crews and officers deserted. Also the rise of the cotton industry from about 1850 to 1875 in New Bedford drew a great deal of capital from the uncertain whale fishery to the more conservative investments in cotton mills, which were successful from the very start. As whaling died out the mills were built up, and it is owing to these same mills that the city was saved from becoming a deserted fishing village. Then later even the lubricating oils began to be made from the residuum of kero-

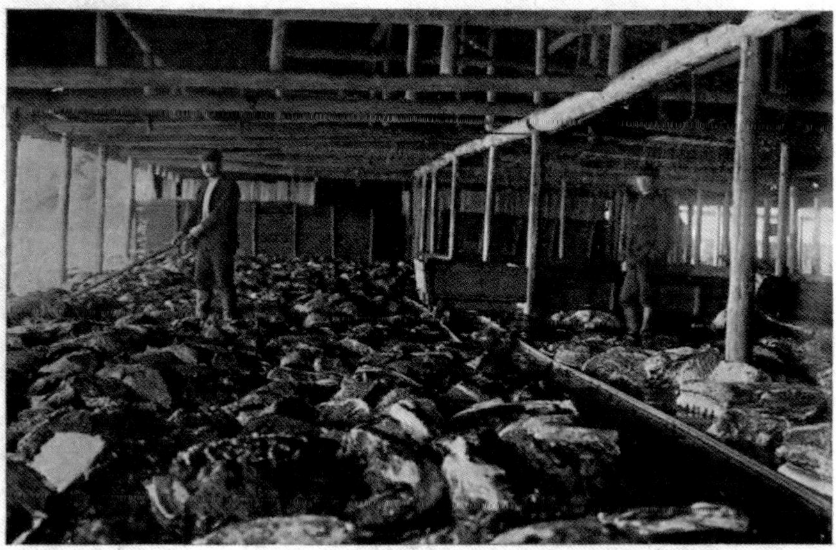

Whale-meat in Japan awaiting shipment to market. It is sold to the poorer classes in all the large towns at prices which range from 7 to 8 cents a pound. One whale yields as much meat as a herd of 100 cattle.

sene, and about the same time wax was invented for candles, which again robbed the whaling industry of another market for oil. Soon came the Civil War, in which many vessels were captured or destroyed, then followed the sinking of forty or more vessels of the Charleston Stone Fleet described elsewhere, and finally came the Arctic disasters of 1871 and 1876, all of which hastened the end of the industry.

WHALING OF TO-DAY

Whaling will doubtless be carried on from San Francisco in a small way as long as there is any demand for whalebone, and from New Bedford and Provincetown while there is any market for sperm and whale oil. Most of the Pacific steam whalers are now provided with a harpoon gun invented by Svend Foyn, a Norwegian. This gun is placed in the bow, and to the harpoon is attached a rope with which to play the whale, as one does a fish with a rod and reel, but there is little romance in this method of whaling.

In modern whaling the flesh is made into guano and the bones and blood into fertilizer, and even the water in which the blubber has been "tried out" is used in making glue. The meat is to-day sold to Japan, and, if the weather is very cold and the supply of fish is limited, a whale might bring there as much as four thousand dollars by utilizing all the by-products as well as the meat, which is sometimes canned. In America a whale is now valued at about two hundred dollars, but, if the entire carcass is utilized, it might bring one thousand dollars.

From the *Whalemen's Shipping List*, still published in New Bedford, it can be figured that the total whaling fleet in America last year (1913) consisted of thirty-four vessels, twenty hailing from New Bedford, eleven from San Francisco, two from Provincetown, and one from Stamford, Conn. The Atlantic fleet, however, reported a total catch of over twenty thousand barrels of sperm oil and one thousand pounds of whalebone during the year 1913, which is a considerably larger amount than for the year previous.

Whaling in stout wooden ships on the far seas of the East and the West is no longer carried on, for the glory and the profit of the industry have gone never to return. Substitute products have come in, and to-day the little whaling that is still done is along the coasts of the Antarctic and Arctic Oceans, off the shores of Western Africa, Northern Japan, New Zealand, California, and South America, and in the main it is carried on in stout iron steamers. Ere long the last whaleship will disappear from the sea and only the romance of a great industry will remain.

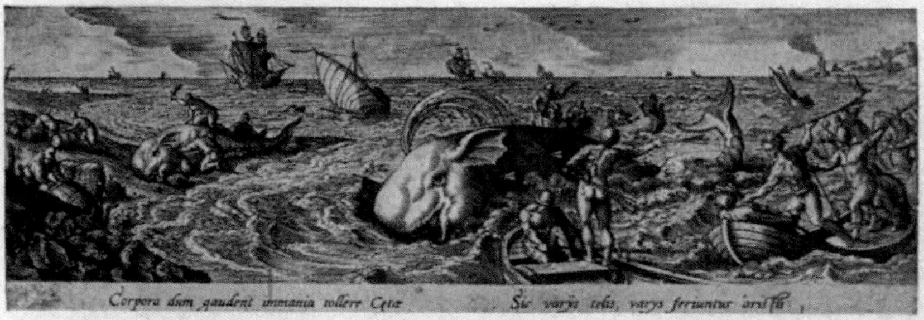

A very old picture of whale-killing in the 17th century.